1005103427

THE SCIENCES IN THE EUROPEAN PERIPHERY
DURING THE ENLIGHTENMENT

Archimedes

NEW STUDIES IN THE HISTORY AND PHILOSOPHY OF SCIENCE AND TECHNOLOGY

VOLUME 2

EDITOR

JED Z. BUCHWALD, *Bern Dibner Professor of the History of Science at MIT*, and *Director of The Dibner Institute for the History of Science and Technology, Cambridge, MA, USA.*

ADVISORY BOARD

HENK BOS, *University of Utrecht*
MORDECHAI FEINGOLD, *Virginia Polytechnic Institute*
ALLAN D. FRANKLIN, *University of Colorado at Boulder*
KOSTAS GAVROGLU, *National Technical University of Athens*
ANTHONY GRAFTON, *Princeton University*
FREDERIC L. HOLMES, *Yale University*
PAUL HOYNINGEN-HUENE, *University of Konstanz*
EVELYN FOX KELLER, *MIT*
TREVOR LEVERE, *University of Toronto*
JESPER LÜTZEN, *Copenhagen University*
WILLIAM NEWMAN, *Harvard University*
JÜRGEN RENN, *Max-Planck-Institut für Wissenschaftsgeschichte*
ALAN SHAPIRO, *University of Minnesota*
NANCY SIRAISI, *Hunter College of the City University of New York*
MERRITT ROE SMITH, *MIT*
NOEL SWERDLOW, *University of Chicago*

Archimedes has three fundamental goals; to further the integration of the histories of science and technology with one another: to investigate the technical, social and practical histories of specific developments in science and technology; and finally, where possible and desirable, to bring the histories of science and technology into closer contact with the philosophy of science. To these ends, each volume will have its own theme and title and will be planned by one or more members of the Advisory Board in consultation with the editor. Although the volumes have specific themes, the series itself will not be limited to one or even to a few particular areas. Its subjects include any of the sciences, ranging from biology through physics, all aspects of technology, broadly construed, as well as historically-engaged philosophy of science or technology. Taken as a whole, *Archimedes* will be of interest to historians, philosophers, and scientists, as well as to those in business and industry who seek to understand how science and industry have come to be so strongly linked.

Archimedes

2
New Studies in the History and Philosophy of Science and Technology

The Sciences in the European Periphery During the Enlightenment

edited by

KOSTAS GAVROGLU

University of Athens, Athens, Greece

KLUWER ACADEMIC PUBLISHERS
DORDRECHT / BOSTON / LONDON

A C.I.P. Catalogue record for this book is available from the Library of Congress.

ISBN 0-7923-5548-2

Published by Kluwer Academic Publishers,
P.O. Box 17, 3300 AA Dordrecht, The Netherlands.

Sold and distributed in North, Central and South America
by Kluwer Academic Publishers,
101 Philip Drive, Norwell, MA 02061, U.S.A.

In all other countries, sold and distributed
by Kluwer Academic Publishers,
P.O. Box 322, 3300 AH Dordrecht, The Netherlands.

Printed on acid-free paper

All Rights Reserved
© 1999 Kluwer Academic Publishers
No part of the material protected by this copyright notice may be reproduced or
utilized in any form or by any means, electronic or mechanical,
including photocopying, recording or by any information storage and
retrieval system, without written permission from the copyright owner.

Printed in the Netherlands

CONTENTS

PREFACE vii

ANA SIMÕES, ANA CARNEIRO, MARIA PAULA DIOGO / Constructing Knowledge: Eighteenth-Century Portugal and the New Sciences .. 1

DIMITRIS DIALETIS, KOSTAS GAVROGLU, MANOLIS PATINIOTIS / The Sciences in the Greek Speaking Regions during the 17th and 18th Centuries 41

AGUSTÍ NIETO-GALAN / The Images of Science in Modern Spain 73

LUIGI CERRUTI / Dante's Bones 95

ARNE HESSENBRUCH / The Spread of Precision Measurement in Scandinavia 1660-1800 179

PREFACE

This issue of *ARCHIMEDES* concerns a subject often referred to as "reception studies". The articles that follow examine particular cases of 'reception' in ways that emphasize pressing historiographical and methodological issues. Such issues arise in any consideration of the transmission and appropriation of scientific concepts and practices that originated in the several 'centers' of European learning and, then, appeared (often in considerably altered guise) in regions of the European 'periphery'.

Although some important work has been done on the subject, themes surrounding the transfer of new scientific ideas, the mechanisms of their introduction, and the processes of their appropriation at the periphery (including the Scandinavian countries, the Iberian Peninsula, the Balkans, and Russia) have not been studied systematically. Many such themes naturally suggest themselves. Here we will single out seven for particular notice: the ways in which the ideas of the Scientific Revolution were introduced to these countries, the particularities of their expression in each place, the specific forms of resistance encountered by these new ideas, the extent to which such expressions and resistances displayed national characteristics, the procedures through which new ways of dealing with nature were made legitimate, and, finally, the commonalities and the differences between the methods developed by scholars at the 'periphery' for handling scientific issues and those of their colleagues in the 'central' countries of Western Europe. These themes, as well as others, will frame our discussions of the complex relationship between "the original ideas of the center" and their "reception in the periphery". The articles in this volume examine a number of these issues in Portugal, Greece, Spain, Italy and the Scandinavian countries. There are, unavoidably, omissions and articles about Russia, other countries of the Balkans except Greece and, most importantly, about the Ottoman Empire would have been absolutely essential complements to these.

Although a simple bipolar distinction between center and periphery is useful for broadly delineating the situation, it is incapable of capturing many salient details. Three factors in particular require expanding any static, bipolar conception. There are, first of all, many centers and many peripheries. Moreover - depending on the subject one is discussing - a place may at one and the same time be both center and periphery. A center may, over time, change into a periphery, and vice-versa. And a single country may contain both centers and peripheries, thereby making purely national distinctions of dubious use. To examine such issues requires discussing the ways in which ideas that originate in a specific cultural and historical setting are introduced into a different milieu with its own intellectual traditions as well as distinctive political and educational institutions. Any such discussion should rather emphasize the 'appropriation' than the unadorned 'transfer' of ideas. Although the concept of idea transfer can be useful and even fruitful for further research, one must always recognize that ideas are not simply transferred like, as it were, material commodities. They are always transformed in unexpected and sometimes startling ways as they are appropriated within the multiple cultural traditions of a specific society during a particular period of its history. Indeed, a major challenge for historians who examine processes of appropriation across boundaries is precisely to transcend the merely geographical, and to concentrate especially on the character of, what one might call, the "receiving culture".

Europe is presently in the throes of its most dramatic transformations since the end of the Second World War. New nations states come into being, new borders emerge, new institutions appear, and old institutions restructure themselves. Many historians and other scholars will look again at the past in the light of current changes. The work that has already been done, as well as newly available sources, combine with (comparatively) open intellectual environments and increases in funding for trans-national contacts to offer an unprecedented opportunity for a critical re-examination of the historical character of European science.

One of the most intriguing challenges for historians of science, technology and medicine is to chart their own thematic atlas within this geographically expanded and culturally diverse Europe, whose present configuration provides a unique opportunity for symbiosis between established and emerging communities of historians. Members of newer communities will, hopefully, decide how to recast what have often, and for many years, been local topics in ways that can be linked to contemporary historiography of science.

Some historians in the emerging scholarly communities will certainly feel that the present moment is an opportune time to, as it were, set the record straight in respect to national contributions. This desire to bring justice to what many may consider to be their misunderstood past is, of course, a natural one and may lead to insistence on the production of biographies of scientists who, it is felt, have been unduly or unfairly neglected. Many such biographies will be interesting and significant by any standards, provided that they are not undertaken solely in the service of a local agenda.

Nevertheless, that sort of agenda must be resisted, not only because it is (at best) purely parochial, but also because in contemporary Europe historians are at a fortunate juncture that offers an unprecedented, and perhaps fleeting, opportunity to expand the domain of problems and issues in the history of science. Consider, as only one among many such opportunities, the issues that arise in considering the European periphery during the Enlightenment. What do we mean when (as many of us do) we write here of Danish, Italian, Greek, Portuguese or Spanish science instead of the sciences as practiced in Denmark, Italy, Greece, Portugal, or Spain? Should we treat this terminological shift as one from geography to culture? Appropriation rather than transmission may here provide a much more satisfying and finely detailed account of the history of scientific practice during this era and in these places, and any such account will require an expansive view that moves beyond purely local agendas.

The following questions point to some of the issues raised by the present volume. What has been the role of new scientific ideas, texts and popular scientific writings in forming the rhetoric concerning modernization and national identity? What scientific institutions became prevalent as power was consolidated and opposition by local scholars emerged? What were the characteristics of the prevailing mode of scientific discourse among local scholars? What was the relation between political power and scientific culture? What were the social agendas, educational policies and (in certain loci) the research policies of scientists and scholars? What shifts in ideological and political allegiances were brought about as the landscape of social hierarchy changed? What consensus and tensions appeared as disciplinary boundaries formed, especially as reflected in the establishment of new University chairs? Finally, what ideological undertones characterized the disputes, and what was their cognitive content?

Ana Simoes, Ana Carneiro and Maria Paula Diogo in their article *Constructing Knowledge: Eighteenth Century Portugal and the New Science,* present the introduction, dissemination and consolidation of the Scientific Revolution in Portugal through the contributions in the 18th century of the *estrangeirados*, an informal network composed of Portuguese who for various reasons were in contact with European intellectual circles, and of the foreigners who had established themselves in Portugal. In the first half of the 18th century, the fate of the Scientific Revolution relied primarily on the endorsement of its ideals by individual personalities, mainly dilettanti and polymaths, who propagated the new ideas through broad but mainly private discussion sessions restricted to an Enlightened elite, and with the translation into Portuguese of some landmarks of the new sciences. A different situation arose during the reign of King José I, when Enlightenment ideas were embodied in legal and administrative measures of which the reform of the University of Coimbra (1772) became paradigmatic. The first textbooks written in Portuguese, up-to-date accounts of science, and critical reappraisals were published, addressing a well-defined audience of students and fellow scientists. The scientific dimension of the new discourse showed a strong emphasis on the qualitative aspects of experimentation, and on the applications of science to potentially useful ends. This utilitarian approach to science was to become a constitutive dimension of Portuguese science itself. As it is typical in peripheral countries, the emphasis was not on production of knowledge but rather on reproduction and propagation of novelty.

Dimitris Dialetis, Kostas Gavroglu and Manolis Patiniotis in their *The Sciences in the Greek Speaking Regions During the 17th and 18th Centuries. The process of appropriation and the dynamics of reception and resistance*, discuss the introduction of the new scientific ideas during the Enlightenment in the Greek speaking regions of the Ottoman Empire. They argue in favor of abandoning the notion of «transfer» and adopting the notion of «appropriation» for the reading of the developments of this period. Their main conclusion is that the Greek scholars who introduced the new scientific ideas developed an idiosyncratic philosophical discourse which was the confluence of ancient Greek philosophy, Eastern Orthodox Christian theology and aspects of the newly emerging scientific discourse in Europe. A synthesis of elements of ancient Greek thought with Orthodox Christian tradition had already emerged by the 18th century as a strong cohesive element in the intellectual identity of the Greek nation; the legitimation of the new scientific ideas ran parallel with economic and political restructuring, both assisting in the formation of a new coherent ideology and political stand, connecting the past of the Greeks with their future prospects as independent nation. Some of the standard scientific texts written in Greek during the 18th century are, also, examined and it is suggested that the reason for the unwillingness of the scholars to initiate breaks with ancient philosophy and theology are to be sought in their overall agendas where political considerations and, especially, issues about national consciousness, were rather prominent.

Agusti Nieto-Galan in his *The Images of Science in Modern Spain. Rethinking the 'Polemica'* comments on some of the most relevant episodes of that long-standing public controversy, commonly known today as "la polemica de la ciencia espanola". He discusses the effect that a frequent negative image of a 'weak' Spanish Science, and the resulting passionate reaction of national pride has had on the local historians of science and technology and, thus, a review of the 'polemica' may contribute to a reassessment of some traditional historiographical problems, and to a fuller understanding of the role of science and technology and their public image in Spain. Following up the 'polemica', and tracing back some episodes of the controversy about the nature of Spanish science and its image among scientists, writers and intellectuals, the paper analyzes how a tacit and recurrent inferiority complex that the Spanish felt and expressed vis-a-vis Europe influenced the scientific debate itself, and as a result, shaped the way in which the history of Spanish science has been written and transmitted to younger generations. Thus Spanish historians of science have often constructed their narratives stressing counter arguments, through a diffusionist model, accepting too easily a view of Spanish science as a mere imposition of a dominant scientific culture from the North, partially neglecting the study of the plurality of sites for creating and reproducing scientific knowledge.

Luigi Cerruti in his *Dante's Bones.Geography and History of Italian Science, 1748-1870* starts his paper with a number of historiographical comments concerning the notion of «Italian» as it has been used in the study of the scientific community. The second part of the paper discusses some aspects of the relationship between political power and scientific culture in Italy of mainly the 18th and 19th centuries, since such a relationship has always been a sensitive issue in the study of Italian culture. A number of questions in the history of Italian science are, then, analysed: the establishment of a 'national' academy, the start of 'national' meetings of scientists, the considerable tradition in astronomical research, and the transition from eclectic local journals to 'national' and specialised ones. There is, finally, a review of the past and present approach of Italian historiography on the way Italian science is located in the context of 'science as such', or as regards to the international 'scientific centres'. The author's conclusion is that if one looks only to the markets for goods, training and information that characterized the activities of some specialities, then Italy was a region in the periphery of these markets. If due attention is paid to the variety of scientific disciplines and specialities, then the differentiation between centre and periphery is no longer applicable in the case of Italy.

Arne Hessenbruch in his article *The Spread of Precision Measurement in Scandinavia 1660-1800* argues that in contrast with the Southern European periphery (the Greek part of the Ottoman empire, Portugal, Spain), 18th-century Scandinavian states resembled the European centre in that they expanded their administrative machinery greatly - to a large extent in order to pay for many expensive wars. One aspect of this development was the establishment of a machinery for raising taxes. This involved precision measurement: surveyors measured land which was enclosed and privatised, and most kinds of merchandise (such as grain or alcohol) were routinely evaluated at town gates for the purpose of levying a duty. Much of the activity of the national Academies of Science was aimed at providing solutions for such quantifying demands within the fiscal system.

Many of the arguments in all the papers have been further clarified and sharpened as a result of extensive linguistic interventions and other substantial comments by Jehane Kuhn. We thank her very much.

Jed Buchwald
Kostas Gavroglu

ANA SIMÕES*, ANA CARNEIRO**, MARIA PAULA DIOGO**

CONSTRUCTING KNOWLEDGE: EIGHTEENTH-CENTURY PORTUGAL AND THE NEW SCIENCES

1. THE PORTUGUESE HISTORICAL CONTEXT IN THE 18TH CENTURY

The high point of Portuguese science is associated with the Portuguese geographical discoveries of the fifteenth and sixteenth centuries. A period of decline ensued, which according to some historians continued until the middle of the eighteenth century. The revival of science associated with the modernisation of the Portuguese economy was linked to two events: the 1772 reform of the University of Coimbra implemented by the Marquis of Pombal (1699-1782) and the creation of the Royal Academy of Sciences of Lisbon in 1779, during the reign of Queen Maria I. Long before Pombal's reforms, however, there was an awareness of the "new sciences." In the final decades of the seventeenth century and the first half of the eighteenth century, some Portuguese showed interest in modern ideas, thereby starting a process which came to full fruition in the second half of the eighteenth century. The introduction, dissemination and consolidation of the Scientific Revolution in Portugal took a considerable time; it extended throughout the eighteenth century, covering the reigns of King João V (1707-1750), King José I (1750-1777) and Queen Maria I (1777-1792).[1] Although they all shared an Absolutist political framework, there were considerable fluctuations in the political and religious orientations of the different monarchs.

The reign of King João V was strongly marked by the abundance of gold coming from Brazil. The availability and circulation of great quantities of this precious metal was to shape particular social and economic structures, diplomatic options, and cultural postures and practices. Immediately after his accession to power, João V had to face a complex network of European allegiances which had been constructed in response to the War of Succession in Spain (1701-1714). Portuguese involvement in that war had already begun with the support of King Pedro II (1683-1706) for French claims. The King was later to favour an alliance with England and Holland. The diplomatic and frequent shifts of allegiance among the European great powers, in particular, England, Holland, Spain, France and Austria, led King João V, and his adviser Cardinal da Mota,[2] to adopt a fundamentally Atlantic external policy, based on trade, on economic autonomy thanks to Brazilian gold, on the exploration of the strategic potential of Portugal's geographic situation; and on neutrality in foreign affairs.

Internally, the exploitation of Brazilian gold produced an increase in profits and an apparent recovery of the Portuguese economy. This resulted in a dynamic market and in a short-lived industrial development led by Luís de Menezes (1632-1690), 3rd Count of Ericeira. He began by reorganising and upgrading industries in order to create a national industrial infrastructure and then proceeded to create new textile plants at Fundão, Covilhã and Portalegre. This industrial strategy was also carried on by some individual initiatives. However, the illicit trade in Brazilian gold, the difficulty of controlling Atlantic routes, the problems arising from the management of the Oriental Portuguese Empire (1736-1740), and the lack of an investment policy soon led to a situation in which the traditional economic weakness had to be confronted again.

Moreover, there were growing symptoms of political and social instability. In particular, the insubordination of the aristocracy in opposition to explicit moves tending to centralise the monarch's power (1728);[3] troubles with the Papal Court (1728-1737);[4] labour conflicts such as the masons' strike at the Convent of Mafra's works (1731) and the "siesta campaign" over the establishment of a lunch-break for workers; religious quarrels such as the anti-semitic persecutions (1730-1735); and the refusal of the religious orders to comply with norms designed to bring them under central control, all these factors added to the crisis.

Culturally and intellectually, the reign of King João V was a contradictory period. On the one hand, the Inquisition's power was reinforced, on the other, the diffusion of the prerogatives of the clergy over the teaching system and its control over culture were increasingly criticized.

The reign of King José I was marked by a clear attempt to modernise Portugal, at political, administrative, economic and cultural levels. In effect, economic growth had been based almost exclusively on the availability of great quantities of gold from Brazil. This caused a hypertrophy of the circulation structure, at the expense of an increase in manufacture.

Although José I was an Absolutist monarch, and an advocate of enlightened despotism, he was followed closely and often guided in his policies by Sebastião José de Carvalho e Melo, the Marquis of Pombal. A former diplomat, Pombal[5] enforced political measures in the framework of Enlightened despotism which aimed at bringing Portugal to the group of modern European nations. Opportunities for industry were increased and state power was increased, in response to an acute economical crisis. The unstable economical situation latent throughout the eighteenth century was worsened by the Brazilian gold crisis of 1766-1767, together with decreasing trade in sugar, diamonds and slaves. The profits of the Companies of Grão-Pará and Maranhão (Brazil) fell drastically and became negligible, in 1770. Especially from 1770 onwards, the Marquis of Pombal enforced measures for the implementation of industry, aimed at reducing importations and restoring the balance of trade. In addition, he launched a commercial policy based on the establishment of state-controlled monopolist companies. These measures were intended to protect the Crown and privileged social groups, in particular, civil servants, great merchants and the Court nobility.

These economic policies were accompanied by a reorganisation of the tax system, which obviously, was to have impact on political power. The provincial and overseas aristocracy, as well as the Jesuits,[6] who were highly influential and owned important sources of income, strongly opposed the centralisation of power. As a result, the Jesuits became one of the main targets of Pombal's persecutions: their teaching activities were forbidden, their properties confiscated, and finally they were expelled from the country, in 1759. Regarding the aristocracy, the persecutions directed at the Houses of Aveiro and Távora, whose members were almost completely exterminated, became famous.

Pombal's reform of the teaching system was part of a plan of attack aimed at reducing clerical power. The Oratorians,[7] a religious order which was the main educational alternative to the Jesuits, had their activities reduced. Their influence was replaced by the lay presence of the state. All these policies were shaped by an ideology of scientific rationalism based on the Enlightenment. Their implementation was supported by the creation of the Colégio Real dos Nobres (Royal College of Nobles) (1761-1837) and the reform of the University of Coimbra (1772). The College of Nobles was intended to create an enlightened nobility capable of participating actively in the modernisation of the country. The great novelty introduced by the reform of the University of Coimbra was to provide an institutional framework within which men of science could carry out their activities on a professional basis; a research infrastructure was put in place, and original investigation was required.

The reign of Queen Maria I[8] may be defined as an attempt to continue the modernisation effort initiated during the reign of her father King José I, while avoiding radical and controversial measures like those implemented by the Marquis of Pombal. Governing policies tended to be moderate, and relations with the aristocracy and the Jesuits were restored at the political level.

From the economic point of view, commercial freedom was favoured; at the same time, industries and the re-structuring of those in existence were encouraged. Favoured by the international situation,[9] this industrial development allowed considerable commercial growth.

From the scientific point of view, the Queen authorized the foundation of the Academia Real das Ciências de Lisboa (Royal Academy of Sciences of Lisbon), by the end of 1779. This was a further fundamental step in the establishment and organisation of a Portuguese scientific community.

Maria I was also to protect educational institutions such as the Aula Pública de Debuxo e de Desenho (Public School of Sketching and Drawing) at Oporto; the Aula Régia de Desenho (Royal Drawing School); the Academia do Nu (Academy of Life Drawing); the Academia Real da Marinha (Royal Naval Academy); and the Academia Real das Fortificações, Artilharia e Desenho (Royal Academy of Fortifications, Artillery and Drawing) all established in Lisbon. She also ordered the foundation of the Royal Public Library.

Together with Pina Manique, her Superintendent of Police, Maria I also launched a wide-ranging group of measures aimed at controlling criminality in Lisbon. Pina Manique (1733-1805)[10] worked towards the neutralisation and control of marginal groups operating in Lisbon, especially prostitutes, thieves and tramps. His intention was to convert the Portuguese capital into a modern and civilised city, following European models. Within this context, the Casa Pia de Lisboa,[11] the first Portuguese technical school aiming at the rehabilitation of the destitute, was set up in 1782.

2. DEFINING THE HISTORICAL QUESTIONS

This paper focusses on the introduction, dissemination and propagation of the "new sciences" in 18th century Portuguese mainland. Although the specific characteristics of the reception of science in various national contexts is a lively topic,[12] the international community of historians of science has scarcely addressed the history of science in Portugal. There is no study of the subject per se, and in the most recent work in which Portugal appears, it is in the context of the history of science in the Iberian Peninsula as a whole.[13] It is misleading to conflate the Portuguese and the Spanish cases; the claim can be made that there is no received view on the topic of Portuguese science at the level of the international scholarly community.

In Portugal itself the situation is quite different. Although a professional community of historians of science has been building up during the last decade, the study of the history of science in 18th century Portugal has been pursued mainly in the context of cultural history. Sporadic contributions in the "évenementielle" genre have also been made by historians and by scientists.[14] However, a consistent study of eighteenth-century Portuguese science was carried out over about forty years by the physics teacher, pedagogue, poet, and historian of science Rómulo de Carvalho (1906-1997). Rómulo de Carvalho belonged to a descriptive tradition that has produced many detailed and highly erudite accounts. He contributed case studies, general works and various papers[15] taking a somewhat Whiggish approach to the eighteenth-century Portuguese scientific context. He focussed almost exclusively on lasting contributions to scientific disciplines such as physics, emphasizing the role of experiment, astronomy and the natural sciences, and taking for granted present-day disciplinary boundaries.

In this paper a different methodological approach is taken. The reception of science in a peripheral country is the object of our study. Despite the European geographic situation of Portugal and the contributions emerging from the 15th century Portuguese discoveries to the process leading to the Scientific Revolution, by the 18th century Portugal was a peripheral country both geographically and scientifically. How were the "new sciences" introduced in Portugal? By whom? What legitimized the new scientific discourse? Was the reception process smooth? What kinds of resistances can be identified?

3. AT THE CROSSROAD: ARCHAISM VERSUS MODERNITY. THE REIGN OF KING JOÃO V (1707-1750)

Especially in the early eighteenth-century, conditions favorable to the concepts of progress and modernity are recognisable in Portugal. Portuguese society was experiencing the beginning of a confrontation between the archaic structures of an Empire locked in its own economic inertia, and the desire to build social structures organised around the ideology of progress.[16] At the social and economic level, this ideology presupposed a transformation of the productive structure; culturally it was characterised by adoption of concepts of scientific and technological rationality.

During the sixteenth and part of the seventeenth-century, the Universities of Coimbra and Évora, both controlled by the Jesuits, had tried to adapt to new European intellectual trends through a re-structuring of Aristotelianism and Scholasticism. In the framework of neo-Aristotelianism, work of the "School of Coimbra" focussed on the publication of commentaries on Aristotle and Thomas Aquinas and culminated in a series of volumes entitled *Cursus Conimbricensis*, which had such an impact that it was adopted by other Jesuit colleges throughout Europe, and praised by many including Descartes.

In Portugal, the Company of Jesus remained attached to very traditional patterns throughout the seventeenth and eighteenth-century. However, examples of more open-minded Jesuits can be singled out. For example, the Italian Christophoro Borri,[17] known in Portugal as Cristóvão Bruno, taught the Copernican theory and mentioned the astronomical discoveries of Galileo in his Aula da Esfera (Course on the Sphere) at the College of Santo Antão, Lisbon. The course was taught mainly by foreign teachers and Bruno, in particular, presented the Copernican theory in a clear and accessible way while, officially, he denied its truth.

In every country at every period, demands for modernization vary, since "modernity" is a particular reaction against a specific state of affairs. In Portugal modernity was defined loosely as everything which opposed Aristotelianism and Scholasticism. In the final years of the seventeenth century, the growing emphasis on scientific and technological research defined a new epistemological framework that called for a radical change in the relation between theory and practice. The reinterpretation of Aristotle was insufficient, and a new philosophy, a new physics and a new methodology were required.

The practice of experiment as a crucial element in the new epistemological context was to have one of its strongest pillars in the Oratorians. The Oratorian João Baptista (1705-1761) was among those who fought Scholasticism, attempted to rehabilitate Aristotle by showing how his texts had been generally distorted by Portuguese commentators, and reconstructed what he considered to be the genuine Aristotelian philosophy in his book *Philosophia Aristotelica Restituta* (1748)[18] in which Baptista reconciled Aristotelianism with Newtonian physics. From 1737, he became the most influential Oratorian advocating experimental philosophy in the Court of King João V.

The Oratorians were endowed by João V with an appropriate building, the Casa das Necessidades, in which they established an open library, a cabinet for natural sciences equipped with appropriate apparatus, and a printing-office. In 1750, they ran a course on experimental physics which, by advocating experiment in the context of the confrontation between the ancients and the moderns, became the centre of a lively controversy. In the satirical pamphlet *Mercurio Philosophico* which circulated in Lisbon, this course was ridiculed:

> The demonstration of a pneumatic machine and the apparent resurrection of a rabbit had just ended, leaving the gentlemen of the audience astonished. Many were the comments on the demonstration (...) I heard one person asking if in the time of Aristotle a similar machine ever existed. As the reply was negative, he retorted:
>
> - Do we know already more than Aristotle?
>
> - Yes, answered the priest demonstrator. He added that in physical matters Aristotle had been rather sluggish.[19]

During the reign of João V, the Oratorians competed with the Jesuits for supremacy over the teaching system; they were to impress their mark upon various intellectuals and men of science, who were either educated by them or belonged to their congregation. The dispute between Jesuits and Oratorians over the control of the Portuguese teaching system is emblematic of the process of dissemination of the Enlightenment in Portugal. However, the main characters of this process were the *estrangeirados* (Europe-oriented intellectuals), who more vividly represented the new spirit of the Scientific Revolution. The *estrangeirados* constituted an informal network composed both of resident foreigners and Portuguese, who had established contacts with European intellectual circles, either through academic education abroad or in their professional life.[20] The *estrangeirados*, diverse in their background and careers, were linked by their wish to introduce into Portugal the scientific rationality of the Enlightenment, and by opposition to the Inquisition and the power of religious fanaticism. (See appendix at the end of the paper identifying key figures in the dissemination of the new sciences in 18th century Portugal).

Examples of *estrangeirados* advocating a reform of cultural, social, political and economic ideas, were Luís da Cunha (1662-1749),[21] his follower Alexandre de Gusmão, and the Count of Ericeira (1673-1743)[22] together with his circle of friends and *protégés*. In particular, the Ericeira Circle, its activities, debates and publications, constituted a true *avant-garde*. They embodied a new rationality and a new kind of relation between the productive structures of knowledge and the whole of society, and in effect, they re-formulated the discourse and practice of Portuguese intellectuals. At the meetings promoted by the Count of Ericeira, known as "Learned Conferences", "daring subjects" were discussed – questions of method, modern philosophy and logic, and the very notion of progress. Their faith in the superiority of the moderns, in alignment with European philosophers such as Jean Bodin or Joseph Glanvill,[23] is apparent from the following excerpt.

> Undoubtedly, our contemporaries have equal or even greater qualities of judgement and reasoning than their predecessors. This is shown by the perfection and grandeur of today's arts and sciences. The books written by the moderns exceed in number, method, contents, quality and elegance all the work of the ancients. (...) In the industrial arts new useful inventions are regularly produced in order to ease daily life. With telescopes one can see even the most hidden regions of the moon; with microscopes the anatomy of atoms becomes possible; with thermometers the degrees of heat and cold can be measured; with barometers the heaviness and lightness of air can be known (...).[24]

The "Learned Conferences" promoted by the Count of Ericeira, incorporated the activities of the Academia dos Discretos (Academy of the Discreet) and of the Portuguese Academy (1696-1716 and 1717). The purpose of these private academies was the discussion of scientific and historical questions. In turn the Academy of the Discreet had incorporated the Academia dos Generosos (Academy of the Generous) (1649-1693) which had been founded in Lisbon, in the seventeenth-century by António Alvares da Cunha (1626-1690) and continued by his son Luís da Cunha.[25] These informal gatherings were to find an official counterpart in the Academia Real de História (Royal Academy of History) (1720).

Another outcome of this movement of intellectual renovation was the foundation, in 1749, of the Real Academia Médico-Portopolitana (Royal Medical Academy of Oporto). Behind its creation were two physicians, Manuel Gomes de Lima (1727-1806) and Sachetti Barbosa (1714-1774),[26] who associated to their initiative Castro Sarmento (1691-1762), and the Jewish *estrangeirado* Ribeiro Sanches (1699-1783).

Thus, in the first half of the eighteenth-century the main strategy of the Portuguese intellectuals in legitimating a new discourse and a new practice relied on diffusion through informal or formal gatherings, which led to the creation of private or public academies.

Another strategy was the publication of books of philosophical character for a general audience. Rafael Bluteau (1638-1734),[27] one of the active participants in the "Learned Conferences," authored the *Vocabulário de Lingua Portuguesa*[28] (*Vocabulary of Portuguese Language*, 1712-1721) based on the ideas of the Enlightenment and on "modern" authors; the engineer Manuel de Azevedo Fortes (1660-1749),[29] another member of the "Learned Conferences," wrote the *Lógica Racional, Geométrica e Analítica*[30] (*Analytical, Geometrical and Rational Logic*, 1744) which introduced Cartesianism in Portugal in a systematised presentation; the *estrangeirado* and *converso* Jacob de Castro Sarmento, who had been involved in the creation of the Royal Medical Academy of Oporto, introduced the Newtonian theory of tides to a national audience, in 1737, in a book entitled *Theorica Verdadeira Das Mares conforme à Philosophia do incomparável cavalhero Isaac Newton*[31] (*The true theory of tides, according to the philosophy of the incomparable gentleman Isaac Newton*); and finally the *estrangeirado* Luís António Verney (1713-1792) published the widely read *O Verdadeiro Método de Estudar*[32] (*True Method to Study*, 1746), a philosophical work in which the archaic teaching practices of Portuguese universities were criticised and a pedagogical reform based on Enlightenment ideals was advocated, with special emphasis on the importance of experiment and mathematics in physics.

Simultaneously, a concept of an utilitarian science resulted in a growing interest in technology. The knowledge of Nature, associated with the ideas of happiness and progress linked to the science-technology axis, was seen as a means for the exploitation of Nature and for the improvement of life. New technical practices developed, in particular, in the military, milling industry, sawing and draining. In 1742, Bento de Moura Portugal (1702-1766), a polymath of Newtonian sympathies and a member of the Royal Society since 1740, presented to the Royal family two steam-engines based on those conceived by Savery, in which he had introduced modifications so they could work independently. His contribution won words of praise from the British engineer John Smeaton, who published a paper in the *Philosophical Transactions* entitled "An Engine for Raising Water by Fire; being an improvement of Savery's construction to render it capable of working by itself, invented by Mr DE Moura Portugal."[33]

The very definition of the object of engineering, as a field in which science and technology interact, attracted special attention from the intellectual elite, as the work by Manuel Azevedo Fortes, *O Engenheiro Portuguez*[34] (*The Portuguese Engineer*), epitomises. In this book, the concept of engineering is genuinely modern in as much as Fortes rejected the treatment of engineering as merely practical knowledge.

Despite all these signs of a new mentality, its real impact was restricted to an enlightened elite. In effect, its dissemination was far from wide. This lack of communication between intellectuals and the rest of Portuguese society was a consequence of the weakness of the channels for the dissemination of knowledge, especially the absence of a dynamic cultural medium capable of accommodating the new trends. This discontinuity, which prevented the consolidation of knowledge and of scientific and technological practices, was the result of politically oriented scientific measures; it was to hinder a sustained growth of a scientific community which, unable to affirm itself as such, found expression in isolated and singular personalities.

4. ASSERTING MODERNITY: MOMENTS OF DISRUPTION. THE REIGN OF KING JOSÉ I (1750-1777)

During the reign of King José I, Pombal's educational reforms were inspired by Ribeiro Sanches and Friar Manuel do Cenáculo (1724-1814). Ribeiro Sanches, a Freemason physician and pedagogue, was one of the leading figures among the *estrangeirados*. Mainly concerned with educational policy, he firmly defended the secularisation of the teaching system. In 1759, after the expulsion of the Jesuits from Portugal, and in the context of the reform of the Estudos Menores (primary and secondary education) Sanches was encouraged to write what proved to be one of the most important pieces of Portuguese pedagogical literature, *Cartas sobre a Educação da Mocidade*[35] (*Letters about the Education of Youth*, 1760). At the request of the Marquis of Pombal, Sanches was also asked to advise on a reform of medical studies in Portugal; in his *Método para Aprender a Estudar Medicina* (*Method for learning how to study Medicine*, 1763)[36] he rejected Galeno's doctrines, advocating instead the theories of Boerhaave and Harvey. His work was to inspire the reform of the Faculty of Medicine of the University of Coimbra and the writing of its statutes, providing Portuguese medicine with a scientific basis.

The Franciscan Friar Manuel do Cenáculo[37] became aware of the Enlightenment movement while attending the General conference of the Franciscans in Rome in 1750. The atmosphere in Rome was of transformation; Pope Benedict XIV promoted the spreading of modernity and the implementation of teaching reforms. For one year, Cenáculo was able to consolidate his awareness of the Enlightenment movement; he was to incorporate its principles fully in his activities as a reformer and as a pedagogue.

When he returned to Portugal in 1751, Cenáculo published the *Conclusiones de Logicæ* (*Conclusions on Logic*),[38] in which modern ideas are already expressed showing the influence of Brücker's *Historia critica philosophiæ* (1741-1744) and *Institutiones historiæ philosophiæ* (1747),[39] especially in his consideration of history as a propaedeutic to philosophy. The *Conclusions on Logic* is the first wide-ranging official Franciscan essay published in Portugal, applying a modern philosophical orientation to the teaching of logic. In Portugal, history of philosophy had as one of its first expressions the programme presented by Verney in the *True Method to Study* in which the author, before introducing logic, metaphysics and physics, gave a historical account with a clear methodological and propaedeutic purpose. The *True Method to Study* anticipated Verney's further work on this matter and fostered the publication of larger historical compilations which, like Cenáculo's *Conclusiones*, emerged after 1751.[40]

In Cenáculo's subsequent philosophical work the author criticised Descartes but minimized the role of experiment despite his many references to Newton. He preferred a metaphysical approach to the problems of physics, which may be summarised in two points: he adopted a Newtonian point of view regarding the possibility of metaphysical knowledge, and a privileged place was given to mathematics, stemming from the limitations of metaphysics as seen from a Newtonian perspective.

In 1791 Cenáculo published his book *Cuidados Literários* (*Literary Concerns*),[41] a compendium which aimed at giving guidance, especially in the realm of philosophy, to younger generations. Together with a refutation and criticism of Scholasticism the book shows a growing mathematicism of Cartesian inspiration which contrasts with Verney's pedagogic ideas more inclined to Newtonianism.

As a reformer of the Estudos Terceiros (Higher Education), Cenáculo suggested new methods which were to influence various reforms of the teaching system, especially the reform of the University of Coimbra. His innovative pedagogical ideas led the Marquis of Pombal to appoint him to various political and administrative offices devoted to education and culture, he presided over the activities of the Real Mesa Censória (Royal Censorship Committee), the Junta do Subsídio Literário (Literary Fund Committee) and the Junta da Providência Literária (Literary Provision Committee), which together had functions similar to a Ministry of Education.

In spite of Pombal's and Cenáculo's affinities their attitudes to Scholasticism did not coincide. As a reformer and a legislator Pombal's complete rejection of Scholasticism was manifest in public documents as well as in the private sphere. Cenáculo, on the other hand – unlike Verney, whose position was also of complete rejection condemned Scholasticism's over-use of abstraction, but considered both logic and metaphysics a good exercise for students. Cenáculo's catholicism and eclecticism converged, holding him back from the extreme rejection – of Scholasticism which were to result in its complete exclusion from the 1772 Statutes of the University of Coimbra.

Cenáculo was made Bishop of Beja (Southern Portugal) in 1770. After King José's death he founded the first Portuguese museum of natural history, which was open to the public. He is also associated with the foundation of various libraries and with the promotion and up-dating of others already in existence. At Beja, Cenáculo also promoted (c.1790) the training of the so-called "Mestras de Meninas" (teachers for girls), which constituted a first step towards institutionalized education for women; the official establishment of a general policy for the education of women, however, would only occur in 1815.

The Prime-Minister Pombal and Cenáculo emerged as the two main executants of a political reform of mentalities. With Pombal, science, technology and culture were secularized becoming clearly an affair of the state.

Pombal's foundation of the College of Nobles followed the publication of Ribeiro Sanches's *Cartas sobre a Educação da Mocidade* (*Letters about the Education of Youth*). Unlike its foreign counterparts, the College of Nobles was not a military school as Ribeiro Sanches advocated, but a public boarding school for the education of young aristocrats. Some, but not all, of the subjects taught at the College were those proposed by Sanches. Sanches had restricted scientific education to arithmetic, geometry, algebra and trigonometry, but the College of Nobles was more ambitious. Its syllabus encompassed experimental physics and astronomy, for whose teaching it was endowed with appropriate buildings equipped with the best instruments acquired abroad. Sanches had recommended appointing foreign professors to teach both military discipline and modern science. Since the College of Nobles was dissociated from military purposes, the only foreign teachers called in by Pombal were those in charge of the sciences, for which few qualified teachers could be found in Portugal.

The College of Nobles was not a successful enterprise: among the main causes of its failure were financial difficulties, the attitude of Portuguese nobility who antagonised such an educational programme, disciplinary questions, and the inappropriateness of the courses to students who were too young to profit from them. Scientific teaching lasted for 5 years (1766-1772). Although the college was prepared to accommodate 100 boarders, only 24 pupils enrolled during this period, of these only 5 completed the scientific courses, and only one proceeded to higher education in mathematics.[42]

The reform of the University of Coimbra was far more successful. It embodied a re-structuring of the former faculties and the allocation of appropriate spaces for research.[43] An independent Faculty of Mathematics was created, and the Faculty of Philosophy included new disciplines such as physics, chemistry, mineralogy, zoology, botany and agriculture. The Faculty of Mathematics was endowed with an observatory and the Faculty of Philosophy was equipped with a laboratory of chemistry, a cabinet of physics, a museum of natural history and a botanical garden.

Pombal also planned to create the Congregação Geral das Ciências (General Congregation of Sciences), an institution that was envisioned as an academy of sciences which would gather the most qualified professors of the different faculties and would be a privileged place for the communication and discussion of original research. The General Congregation of Sciences never came to light; Pombal's career came to an end in 1777 when he had not yet approved the statutes of the congregation prepared by the Chancellor Francisco de Lemos (1735-1822).

At the Faculty of Mathematics, algebra, phoronomy (kinematics), astronomy and geometry were taught. The professorships of algebra and astronomy were held respectively by two Italians, Franzini and Ciera, who had been former professors at the College of Nobles. Phoronomy and geometry were taught respectively by the ex-Jesuit Monteiro da Rocha (1734-1819) and the mathematician and poet José Anastácio da Cunha (1744-1787).

At the Faculty of Philosophy the participation of the Italian natural philosophers, Domenico Vandelli (1730-1816),[44] chemist and natural historian, and Dalla Bella (1726-c.1823),[45] physicist, whom Pombal had called in to carry out the reform, had less impact. The task of establishing the Laboratory of Chemistry, the Natural History Museum and the Botanical Garden assigned to these men was far from being successful. Pombal disapproved of Vandelli's and Dalla Bella's plans, considering them too expensive and extravagant. The letter he addressed to the university's chancellor Francisco de Lemos reads in part:

> Those Professors are Italians and the people from that Nation, who are used to seeing hundreds of thousands of Portuguese 'cruzados'[46] being thrown into the air in Rome,[47] are full of enthusiasm and think that everything which is not too expensive does not honour the name of Portugal or their own name.[48]

In the event, both men became more interested in profitable applications of science and Vandelli ended up as a successful entrepreneur of a china factory. He managed to keep his salary from the university, despite his non-compliance with his teaching and research duties. Notably, he was director of the Laboratory of Chemistry and Senior member of the Faculty of Philosophy even after his appointment to various administrative and political positions in Lisbon.

The role of these two Italians was almost negligible in scientific terms; when the German naturalist H. F. Link visited Portugal the following comments were made concerning Vandelli's work at the Botanical Garden of the Ajuda Palace at Lisbon:

> Il ne faut pas s'attendre à rencontrer dans cet établissement une bonne indication des trésors qu'il renferme. Si vous demandez des renseignements, le professeur Vandelly (sic) vous ouvre le *systêma vegetabilium* de Linné (Édition de Murray) et pour peu qu'une description qui s'offre à lui ait quelque trait à la plante en question, ce bota-

niste ne balance pas un instant à lui assigner son nom. Au reste, ce docteur Domingos Vandelly, né en Italie, est connu des naturalistes par quelques ouvrages, mais particulièrement par ses liaisons avec Linné. On ne saurait lui disputer d'avoir été, dans sa jeunesse, un homme studieux, et d'avoir entrepris beaucoup, pour acquérir la célébrité [...] Il est bien arrière pour les connaissances. A peine connaît-il les plantes qu'il a jadis décrites lui-même, il est également mauvais minéralogiste, et ses *Mémoires de Chimie*, insérés dans les Memorias de l'Académie, l'ont couvert de ridicule auprès des savants.[49]

5. ASSERTING MODERNITY: THE ROADS TO CONSOLIDATION. THE REIGN OF QUEEN MARIA I (1777-1792)

Scholarly publication is undoubtedly a powerful means of promoting science among specialists, training professional scientists and at the same time testifying to the liveliness of a scientific community. An effort towards a more consistent and specialized editorial activity was enforced by Pombal and continued throughout the reign of Queen Maria I.

According to the new Statutes of the University of Coimbra, in addition to their teaching and research duties professors were expected to write original textbooks in Portuguese. This measure was not rigorously enforced and, in fact, many professors never complied with it, restricting their pedagogical activitity to the translation of foreign manuals. That was the case of the mathematician Monteiro da Rocha and the naturalist Vandelli. Dalla Bella took a long time to publish a manual entitled *Physices Elementa*[50] (*Elements of Physics*, 1789-1790), which was then criticized for its prolixity and lack of discussion of topics already current in textbooks of the time.

Anastácio da Cunha in contrast, produced an original textbook, *Principios Mathematicos* (*Mathematical Principles*, 1790),[51] which he submitted to the University for approval. Despite the precepts of the statutes, however, no opinion was ever formally expressed and the book was never officially adopted. The *Mathematical Principles* was an original attempt at a formal presentation of mathematical knowledge, in which he established the foundations of mathematical analysis.[52] He gave an original definition of a convergent series which he then applied to the demonstration of the convergence of a geometrical series. He also expounded a new theory of the exponential function based on the respective development in a power series. The book was published posthumously by Cunha's disciples at the Casa Pia, in 1790 in Portugal, and was translated afterwards in French.[53] Despite its minimal impact in the Portuguese milieu of the time, Cunha's contributions were acknowledged by foreign mathematicians such as C. F. Gauss, in 1812, and I. Timtchenko, in 1894[54] (Figure 1).

PRINCIPIOS MATHEMATICOS
PARA INSTRUCÇAÕ
DOS ALUMNOS DO COLLEGIO
DE
SAÕ LUCAS,
DA REAL CASA PIA DO CASTELLO
DE
SAÕ JORGE:
OFFERECIDOS
AO SERENISSIMO SENHOR
D. JOAÕ,
PRINCIPE DO BRAZIL:
COMPOSTOS PELO DOUTOR
JOSE' ANASTACIO DA CUNHA,
DE ORDEM DO DESEMBARGADOR DO PAÇO
DIOGO IGNACIO DE PINA MANIQUE,
Intendente Geral da Policia da Corte, e Reino, &c, &c, &c.
LISBOA
NA OFFIC. DE ANTONIO RODRIGUES GALHARDO,
Impressor do Eminentissimo Senhor Cardeal Patriarca.
ANNO M. DCC. XC.
Com licença da Real Meza da Commissaõ Geral sobre o Exame, e Censura dos Livros.

ELEMENTOS
DE
CHIMICA
OFFERECIDOS
A
SOCIEDADE LITTERARIA
DO RIO DE JANEIRO
para o uso do seu curso de Chimica
POR
VICENTE COELHO
DE SEABRA
Formado em Filosofia pela Universidade de Coimbra &c.
PARTE I.
COIMBRA
NA REAL OFFICINA DA UNIVERSIDADE,
Anno de M.DCCLXXXVIII.
Com licença da Real Mesa da Commissaõ Geral sobre o Exame, e Censura dos Livros.
Foi taixado este Livro com a Dissertaçaõ em [illegible] reis. em papel.

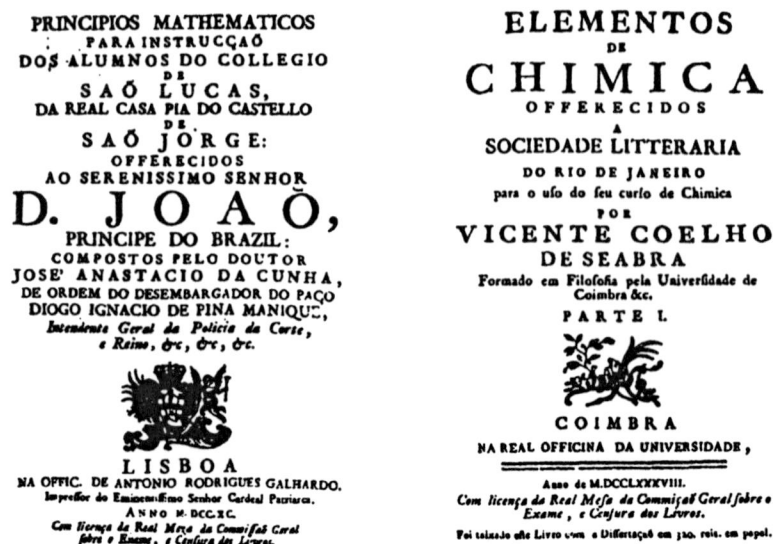

Fig. 1. *Left.* Front page of José Anastácio da Cunha, *Principios mathematicos* (1790), facsimile published by the Department of Mathematics of the University of Coimbra (Coimbra, 1987).

Fig. 2. *Right.* Front page of Vicente Coelho Seabra, *Elementos de Chimica* (1188-1790), facsimile published by the Department of Chemistry of the University of Coimbra (Coimbra, 1985).

The Freemason *estrangeirado* Felix Avellar Brotero (1744-1828),[55] a botanist of international reputation, while still in exile in Paris, had published his *Compêndio de Botânica*[56] (*Compendium of Botany*), which earned him the admiration of his peers. As professor of botany and agriculture at Coimbra since 1790, he reorganised scientifically the botanical garden of the university, wrote a textbook, *Princípios de Agricultura Philosophica*[57] (*Principles of Philosophical Agriculture*, 1793), and later, books devoted to the inventory and study of Portuguese Flora; his *Flora Lusitanica* (1804)[58] and *Phytographia Lusitaniae Selectior* (1816-1827),[59] became landmarks of Portuguese botany.

The introduction of chemistry at the University of Coimbra was to owe a great deal to a Brazilian-born man of science, Vicente Coelho de Seabra (1764-1804).[60] Seabra's textbook *Elementos de Chimica*[61] (*Elements of Chemistry*, 1788-1790) had two parts, the first published one year before Lavoisier's *Traité Elémentaire de Chimie* (1789), when he was still a 24 year old student, and the second part one year after it. Seabra showed a remarkable awareness of the most recent contributions of his fellow European chemists: he had already adopted Lavoisier's theory of oxygen in 1787 the same year that saw the conversion of Guyton de Morveau, Monge, Chaptal and Meusnier while renowned chemists such as Bergman, Scheele, Kirwan, Cavendish and Priestley were still arguing against it.

Seabra's compendium is not an elementary text or a piece of popularisation but an up-to-date and clear presentation of modern chemistry, the first book written in Portuguese in which modern chemistry was introduced to a national audience. Nationally, its impact was almost negligible, despite the approval of the Faculty of Philosophy of the University of Coimbra. In practice, the Faculty continued to use old-fashioned Stahlian textbooks such as Scopoli's *Fundamenta Chemiae*.[62]

Seabra also wrote the *Nomenclatura Chimica Portugueza, Franceza e Latina a que se junta o sistema de characteres chimicos adaptados a esta nomenclatura por Hassenfratz e Adet*[63] *(Chemical Nomenclature in Portuguese, French and Latin...)* (1801), a translation into and an adaptation to Portuguese of Lavoisier's chemical nomenclature. Seabra's chemical terminology was adopted and with slight modifications is still in use (Figure 2).

Another Brazilian-born man of science, the mineralogist José Bonifácio (1763-1838)[64] was to be of paramount importance in the implantation of geology at the reformed university. An original researcher, he discovered twelve new minerals (in fact, four new species and eight varieties of new species)[65] and published his findings in a memoir entitled "A short notice concerning the properties and external characters of some new fossils from Sweden and Norway; together with some Chemical remarks upon the same," first published in the *Allgemeines Journal der Chemie* (1800).[66] This memoir brought him international reputation and was translated and published in France and England.[67] In terms of publications, Bonifácio marks a new phase among Portuguese men of science: his work was published in the form of articles in foreign and national scientific journals rather than in the form of textbooks.

It is clear that after the failed attempt at the establishment of scientific teaching at the College of Nobles, the Marquis of Pombal focussed his attention on the University of Coimbra, believing that the success of this institution could be secured by new and detailed statutes. Non-compliance with many important statutory rules, however, such as the obligation to write original textbooks in Portuguese, their adoption in the lectures, and the production of original scientific research, extended throughout the reign of Queen Maria I: an increasingly heavy bureaucracy, and internal power struggles, combined to prevent renovation and creativity at the university. Many of those holding top positions not only did not fulfill the requirements of the statutes but also managed to convert posts such as director of the Observatory or of the Laboratory of Chemistry into lifelong sinecures, while some fine scholars

Fig. 3. Royal Academy of Sciences of Lisbon (courtesy of Professor A.M. Nunes dos Santos, SHFC/FCT/UNL).

who fulfilled the prescriptions of the statutes, such as Seabra, were kept in junior positions. In the atmosphere of severe political and religious intrigue that followed Pombal's political downfall, Anastácio da Cunha was even forced to leave the university.

Maria I sought to continue the measures of her father, of which the reform of the University of Coimbra became emblematic; the other main contributions to science at the institutional level during her reign were the foundation of the Royal Academy of Sciences of Lisbon, the first Portuguese scientific academy, and of the Casa Pia, the first Portuguese school of technical education.

The creation of the Academy of Sciences was a decisive attempt to establish and organize a coherent scientific community. The first president of the academy was the Duke of Lafões, the secretary was the Viscount of Barbacena (the first person to have completed a doctorate at the newly reformed University of Coimbra), and the vice-secretary was Abbot Correia da Serra (Figure 3).

The Duke of Lafões (1719-1806)[68] represented the enlightened intellectual Freemason, deeply committed to the development and consolidation of rational knowledge. Despite his early interests in philosophy at the University of Coimbra he had studied canon-law. During the reign of King José I, because of his participation in the resistance of the high nobility to Pombal's policies, he was denied his noble title of Duke; in exile he spent much time in London and travelled extensively throughout Europe. He established many contacts with the European intellec-

Fig. 4. Correia da Serra (in *Eulogy of Benjamin Franklin*, by Correia da Serra, courtesy of Luso-American Foundation, Lisboa, 1996).

tual elite, acquired an international reputation, and became a Fellow of the Royal Society before returning to Portugal after Pombal's fall.

Like the Duke of Lafões, José Correia da Serra (1750-1823)[69] epitomised the open-minded and enlightened Freemason of eighteenth-century Portugal. He became a natural historian with a rare wide-ranging culture, and an early advocate of transformist views, making use of the notions of affinity, structure and function. His use of the concept of symmetry was considered very innovative by the renowned Swiss botanist Augustin P. de Candolle.[70] His contributions to various international journals[71] earned him the respect of his international peers such as de Candolle, A. von Humboldt, and Cuvier (Figure 4).

The Duke of Lafões and Correia da Serra, together with the Oratorian priest Teodoro de Almeida (1722-1804)[72] modelled the academy on its European counterparts. Teodoro de Almeida was chosen to read the Opening Address on 4 June 1780, the day the new scientific institution was presented to the public. The address was a long denunciation, not shared by all academicians, against the state in which Pombal's regime had left Portugal. Almeida characterised Portugal as a "centre of ignorance"[73] only comparable to Morocco, and announced that the academy would seek to reverse this state of affairs. A fierce controversy immediately broke out. That conflict has recently been analyzed in terms of alternative ways of envisioning the pathway to modernity: either through Pombal's agenda, particularly the secularization of the teaching system, or by way of benevolent Despotism and Absolutism.[74]

The academy was organised in three classes: the first dealt with the "sciences concerned with observation" and included meteorology, chemistry, anatomy, botany and natural history (its first director was Domingos Vandelli); the second with the "sciences concerned with computation" and included arithmetic, geometry, mechanics and astronomy; the third class with Portuguese language and literature. A physics laboratory was set up and equipped with many of the best instruments of the day. In accordance with the utilitarian view of science advocated by the academicians, committees were created to encourage agricultural and industrial improvements.

Papers submitted to the academy were published in the *Memórias Económicas* (*Memoirs on Economy*) (5 volumes published between 1789 and 1815); the *Memórias de Matemática e Física* (*Memoirs on Mathematics and Physics*) (2 volumes published during the eighteenth-century, 1797 and 1799) and the *Memórias de Literatura Portuguesa* (*Memoirs on Portuguese Literature*) (9 volumes published from 1792 to 1814). The *Memoirs on Mathematics and Physics* included topics ranging from mathematics to natural history, chemistry, physics and astronomy. In mathematics and physics, papers addressed topics then typical among the enlightened, largely related to the understanding, articulation and application of Newtonian physics. Newly developed fields of mathematics were applied to the mathematisation of mechanics and dynamics, and to the geometrisation of space. The memoirs on natural history aimed at giving a new structure to the natural world, one in which man was to have a place in the "chain of being". Chemistry, mineralogy, botany, and particularly the use of the flora of the colonies for medical and pharmaceutical purposes are among the frequent subjects.

The memoirs followed norms guided by what one might call the modern understanding of a scholarly text: they usually contained the definition of the problem in its historical context, followed by a hypothesis and its justification. Quotations were used to legitimate the text, at the same time creating new references for the dissemination of knowledge. Texts were often written in the first person, constructing a subject who conceives experiments, performs and describes them.

The successive alterations introduced in the academy's statutes throughout the eighteenth-century such as lifelong presidency, the creation of the category of veteran members, and the increasing number of honorary associates show that the initial project met with difficulties; the academy became an increasingly formal, less operational institution. On the other hand, the hierarchical organisation of the academy which privileged science over the humanities was not welcomed by a cultural milieu which had always centered on the *belles-lettres*. At the scientific level, the utilitarian agenda of the academicians failed, in the sense that the practical scientific findings delivered in most memoirs were never implemented in projects, that might generate economic wealth and industrial growth.

Pina Manique, Police Superintendent during the reign of Queen Maria I, was a paradoxical character. While he persecuted some academicians on the grounds that they were sympathetic to Freemasonry and French authors such as Voltaire, at the same time he played an active role in the creation of the Casa Pia de Lisboa, a boarding school for destitute adults and children, inspired by the precepts of the renowned Swiss-Italian pedagogue, Pestalozzi. A wide range of subjects was taught according to pupils' capabilities, and various levels of education were covered. The first plan of studies was designed by the mathematician Anastácio da Cunha and aimed to give a solid preparation on both humanities and science. The Casa Pia also provided practical, technical and artistic education and diplomas in surgery and obstetrics, agriculture and veterinary medicine; as part of its utilitarian project the Casa Pia was to supply agriculture and industry with qualified personnel. In addition, it was part of a network of colleges nationwide and abroad (Rome, London, Edinburgh and Copenhagen), with a grant system through which students were sent to the University of Coimbra and foreign universities for post-graduate studies.

The Casa Pia's results were remarkable in quantity and quality, especially in comparison with the College of Nobles or even with the University of Coimbra. It was at the Casa Pia that Anastácio da Cunha was to find most of his disciples,[75] forming a research group of able mathematicians sharing intellectual affinities and a research programme. They were active in disseminating their master's work, and several of them became nationally and internationally recognised. The formation of research groups was a rare event in Portuguese science; more usually, individual scientists were unable to gather around them a group of juniors to carry on the research programme in which they had been trained.

6. ASSERTING MODERNITY: THE POPULARISATION AND DISSEMINATION OF SCIENCE DURING THE REIGNS OF KING JOSÉ I AND QUEEN MARIA I

One of the leading Portuguese in the dissemination of science in part because he published in French and English was the *estrangeirado* João Jacinto de Magalhães (1722-1790),[76] or Jean-Hyacinth Magellan as he became widely known in foreign scientific circles. Through his readings, correspondence and acquaintances, Magalhães was to have an extraordinary impact on scientific communication. Although he made no fundamental contributions of his own, he introduced French men of science to scientific instruments made in England, kept up with the latest developments in English, Scottish, and Swedish chemistry, and introduced Priestley's work to French chemists.

His *Essai sur la nouvelle théorie du feu élémentaire et de la chaleur des corps* (1780),[77] also published in 1781 in the *Journal de Physique*, was an important vehicle for the dissemination of the new theories of heat developed by Black, Irvine and Crawford. This book introduced the term *chaleur spécifique* into scientific literature, and contained the first printed table of specific heats, based on Kirwan's findings; it also described Black's experiments on latent heat.

Magalhães also became interested in pneumatics: he immediately understood the importance of Priestley's work on "airs", which he transmitted to French chemists, and introduced some improvements of his own to Priestley's apparatus for making carbonated water. Another field on which Magalhães left his mark was mineralogy. In 1770, the Swedish mineralogist Cronstedt advocated in the book *Mineralogy* that the mineral kingdom should be studied using the tools of chemical analysis. Magalhães foresaw the importance of the application of these techniques, and suggested to Gustav von Engestrom, a former student of Cronstedt, that he should translate Cronstedt's book into English (1777). He himself began to work on a second English edition,[78] which was considerably enlarged by notes gathered in the course of his correspondence with Cronstedt, Kirwan, Bergman, Guyton de Morveau and Giovanni Fabbroni. Although retaining the old terms of the phlogiston theory, Magalhães could not restrain himself from adding to Cronstedt's text many notes on Lavoisier's new theory of combustion and calcination.

Magalhães's main interest was, however, in the field of scientific instrumentation: astronomical and nautical, meteorological and physical. He wrote books in which he described their functioning and use, with improvements of his own; his modifications of the English quadrant became widely known and adopted. His writings and original contributions to instrumentation may thus be included in the tradition of the British artisan-scientist partnership. His memoirs on scientific instruments were all written in French and published between 1775 and 1783 under the sponsorship of the Académie Royale des Sciences de Paris. As a correspondent member of the Académie, he kept Continental men of science informed about new developments on scientific instrumentation coming from Britain. The contacts Magalhães maintained with his homeland made possible the acquisition abroad of various scientific instruments by the physical laboratory of the College of Nobles, which were later transferred to the reformed University of Coimbra (Figure 5).

During the eighteenth-century, Portuguese literature also contributed to the dissemination of scientific knowledge. Various journals and books devoted to the popularisation of science emerged. Periodical publications were launched such as the *Jornal Enciclopédico* (*Encyclopaedic Journal*, 1779-1793; 1779-1788; 1788-1790) and the *Gazeta Literária* (*Literary Gazette*, 1761-1762).[79] The *Encyclopaedic Journal* focussed on subjects ranging from philosophy, physics, chemistry, medicine, natural history, literature, civil and rural economy to political relations, morals and anecdotes. The journal aimed at encouraging the Portuguese to contribute "useful articles" but in practice extracts from foreign articles and papers written by

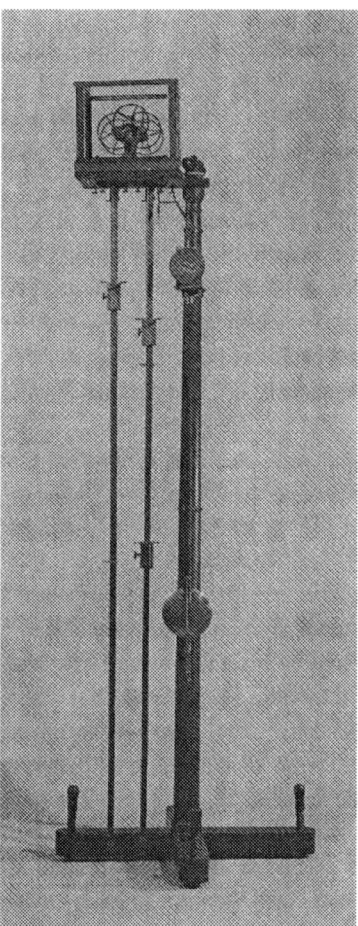

Fig. 5. Atwood machine improved by L.J. Magalhães. Cabinet of Physics of the reformed University of Coimbra (in Catalogue of the Exhibition *O Engenho e a Arte*, Calouste Gulbenkian Foundation, Lisboa, 1997).

the editors came to make up the bulk of its contents. The journal was sold directly to the subscribing public and distributed nationwide; by Portuguese standards its circulation (466) was considerable.

The *Literary Gazette*, despite its name, was a journal devoted to the dissemination of useful science. Its aim was "to shape our judgement and, consequently, our happiness" through the consideration of issues of metaphysics, natural history, astronomy, physics, chemistry, anatomy, law, rhetoric and poetry, and "all those subjects of countless uses to Man's welfare."[80] The list of subscribers covered 200 people, 26 of whom were foreigners, mostly British. The bulk of its contents consisted of critical reviews of the main European publications, rather than national productions.

The life and the contents of these and other journals were at the mercy of political fluctuations, and in particular, of censorship; their publication was interrupted and often completely and definitely suppressed.

Together with the journals popularizing science other forms of literature emerged, which were cultivated especially by writers such as the Marchioness of Alorna, José Agostinho de Macedo, and Teodoro de Almeida. The works of the first two exemplify the genre known as scientific poetry. The Marchioness of Alorna published a poem composed of six cantos entitled *Recreações Botânicas*[81] (*Botanical Recreations*), published posthumously in 1844; Macedo wrote a poem of four cantos entitled *Newton* (1813).

Teodoro de Almeida was one of the most important Portuguese popularisers of the eighteenth-century; his *Recriação Filosófica* (*Philosophical Recreation*),[82] published in 10 volumes from 1751 to 1800, admirably combined depth, breadth, ease of reading and systematization of knowledge.

Addressed to a non-specialised audience, the *Philosophical Recreation* introduced the common reader to modern science in a very accessible and pleasant way. In analogy with Galileo's dialogues, it took the form of a conversation between three characters – the advocate of modern ideas, Teodósio, who actually represented Teodoro de Almeida; the physician Sílvio, who was an advocate of Aristotle, and finally the apprentice Eugénio, who was meant to be converted to modern ideas. The first six volumes, published from 1751 to 1762, were devoted to natural philosophy: the first volume dealt with mechanics, particularly the properties of matter and motion; the second volume discussed phenomena related to the five senses, such as light, colours, sound, heat and cold; the third volume dealt with the four Aristotelian elements; the fourth with optical instruments; the fifth with animals and the sixth with the system of the world. In the last four, Teodoro de Almeida turned away from natural philosophy. Beginning with reflections on historical events such as the French Revolution, these volumes show a preoccupation with the increasing role played by reason in human affairs. They dealt with problems of logic (volume 7), metaphysics (volume 8), the conciliation between reason and religion (volume 9) and ethics (volume 10).

Teodoro de Almeida's *Cartas Fysico-Mathematicas*[83] (*Physical-Mathematical Letters*) were published in 3 volumes from 1784 to 1798 as a supplement to the first six volumes of the *Philosophical Recreation*. In the first volume, addressed to Eugénio, the principles of geometry were laid down in order to prepare him and the reader for the topics of physics discussed in the following volumes: mechanics is the topic of the second volume; the third volume dealt with electrostatics, magnetostatics, the composition of air, and several other rather heterogeneous subjects. Some were memoirs that had been presented to the Academy of Sciences but not published.[84] In fact, it is rather amazing that Almeida, a founding member of the Academy who had been chosen to read the Opening Address, was never able to get his memoirs published in its *Memoirs of Mathematics and Physics*. One reason may have been the widening gap between Almeida's agenda and that of the other academicians.[85] Almeida sought to reconcile natural philosophy with catholicism, an attempt to keep a prominent place for religion in a period when reason was playing an increasingly dominant role. The academicians, on the other hand, advocated an utilitarian approach to science, and promoted social projects directed in part towards the support of agriculture, commerce and industry.

Despite its wider scope the *Philosophical Recreation* has been compared with Benito Feijoo's *Theatro Critico Universal*.[86] The *Philosophical Recreation* and the *Physical–Mathematical Letters* were quite popular in Portugal and Spain, capturing wider audiences than most works published at the time, as shown by the successive revised or re-written editions of some of its volumes. Some of them reached five editions during Almeida's lifetime – a rare event in the Portuguese context.

This was also the case with a novel published in 1779 by Almeida and entitled *O Feliz Independente do Mundo e da Fortuna ou Arte de Viver Contente em Quaisquer Trabalhos da Vida*[87] (*The Happy Independent from the World and Fortune, or the Art of Living Happily through Any Vicissitude of Life*). Almeida's book is a kind of pre-romantic prose epic. In this novel of philosophical allegory and pedagogical purpose, Biblical characters and long moral debates taking place in a naturalistic setting are featured. Whenever necessary, Teodoro de Almeida supports his views with scientific examples, especially drawn from physics and astronomy.

As one of the leading figures of the Portuguese Enlightenment, Teodoro de Almeida advocated a rather idiosyncratic view of modern science. Invited to tell his interlocutors why objects fall, for example, Teodósio replies that it was God who determined that "things separated in a violent way from others, or from Earth, look for Earth by themselves when that violence stops."[88] He adds that a stone does not fall due to Descartes's "subtle matter" or to any attractive property in Gassendi's or Newton's fashion. Almeida later came to endorse Newtonianism; but other examples of eclecticism continued to be found in his works. Teodoro de Almeida's eclectic approach to modern science combined a reaction against the excessive formalism of Scholasticism which had long dominated Portugal with a belief in the provisional status of knowledge, which liberated him from subscribing to one particular philosophical system. Above all, it was a deeply committed attempt at a compromise between different physical systems and the search for theological explanations.

7. CONCLUDING REMARKS

The introduction, dissemination and consolidation of the Scientific Revolution in Portugal took a considerable time. It extended throughout the eighteenth-century, covering the reigns of King João V, King José I and Queen Maria I. The Oratorians, who had been the main rivals of the Jesuits over the control of the teaching system, played a decisive role by cultivating and teaching experimental physics and modern philosophy. Several of the most important men of science were educated by or had strong links to this religious order.

The leading contributors to the diffusion of the Enlightenment in Portugal were, however, the *estrangeirados*, an informal network composed of Portuguese who for various reasons were in contact with European intellectual circles, and foreigners who had established themselves in Portugal. Individually, the *estrangeirados* varied widely in their backgrounds and careers. Some were administrators, diplomats, politicians, physicians, priests or men of science. Some were Jews, *conversos* or Freemasons, who had fled from the Inquisition, or from political persecution.

As a network the *estrangeirados* were linked together by the desire to introduce in Portugal the scientific rationality of the Enlightenment. In the first half of the 18th century, the fate of the Scientific Revolution relied primarily on the endorsement of its ideals by individual personalities, mainly dilettanti and polymaths, who propagated the new ideas through broad but mainly private discussion sessions restricted to an Enlightened elite, and on the translation into Portuguese of some landmarks of the new sciences, using a mainly philosophical discourse from which a group ideology developed. A different situation arose during the reign of King José I, when Enlightenment ideas were embodied in legal and administrative measures of which the reform of the University of Coimbra became paradigmatic. It was in this context that Portuguese men of science were first able to develop a more scientifically-oriented (professional) discourse. The first textbooks written in Portuguese, up-to-date accounts of science, and critical reappraisals were published, addressing a well-defined audience of students and fellow scientists. Subsequently, the publication of the first scientific papers in the Academy of Sciences of Lisbon, and in journals of respected foreign societies, inaugurated in Portugal the era of periodical scholarly publication.

The process of establishing a new mentality was intimately linked to Portuguese politics. In the first half of the eighteenth century, diplomats, administrators and politicians, who actually never pursued a scientific career nevertheless played important roles. In the second half of the eighteenth century, on the other hand, the involvement of men of science in political and administrative spheres, made them vulnerable to political and religious changes, and persecution and exile were frequent. In particular, political interference in academic appointments, led to situations in which even those who did not comply with their duties often kept their posts for life. This misuse of power was to hinder the needed renewal of scientific institutions.

The scientific dimension of the new discourse showed a strong emphasis on the qualitative aspects of experimentation, and on the applications of science to potentially useful ends. On the one hand, this was in part a reaction against the theoretical features of Scholasticism. On the other, it was an attempt to provide technological development with a scientific basis. This utilitarian approach to science was to become a constitutive dimension of Portuguese science itself.

As it is typical in peripheral countries, the emphasis was not on production of knowledge but rather on reproduction and propagation of novelty. Reproduction took as usual an eclectic approach, especially in the period before the institutionalisation of science on a professional basis with the reform of the University of Coimbra and the foundation of the Academy of Sciences. Generally, each Portuguese author took whatever s/he found appealing from foreign sources as the foundation on which to construct a personal interpretation of modern science. This attitude often included theological explanations by-passing what eighteenth-century Portuguese authors found unacceptable in the philosophy of the most representative figures of the Scientific Revolution and the Enlightenment. However, among professional men of science, there were genuinely original contributions. Correia da Serra, Brotero, Bonifácio and Anastácio da Cunha were all innovative in their respective fields, and won a place in an international scientific network.

A mainly illiterate population who still held to superstition and an atavistic religiosity; the difficulty of establishing appropriate channels for the dissemination of science; the failure of national entrepreneurs still attached to traditional agricultural and industrial procedures, to use, promote and apply the scientific research projects of Portuguese men of science, university professors and academicians – all contributed to Portugal's resistance to a widespread consolidation of the conceptual acquisitions of the Scientific Revolution and of the Enlightenment.

Faculdade de Ciências, Universidade de Lisboa, Department of Physics, Portugal
**Faculdade de Ciências e Tecnologia, Universidade Nova de Lisboa, History of Science*

Acknowledgements

Preliminary research to this article was carried out in the context of PROMETHEUS (European project: CHRX-CT93-0299) - The Spreading of the Ideas of the Scientific Revolution from the Countries Where they Originated to the countries in the Periphery of Europe (Iberian Peninsula, Balkans, Scandinavia) during 17th, 18th and 19th centuries, (1994,95,96). A version of this paper was presented at the Meeting "The Transmission of Scientific Ideas to the Countries of the European Periphery, during the Enlightenment," in Delphi, Greece, 23-27 July 1995. We wish to thank the participants for their comments and questions and, in particular, Jed Buchwald, Kostas Gavroglu and Norton Wise for their useful suggestions. Also our gratitude to Francisco Contente Domingues and Kostas Gavroglu, for their careful reading of versions of this paper.

Appendix - Basic identification of key figures involved in the dissemination of the new sciences in 18th century Portugal

Name/Nationality	Profession/Contributions	Educational Background	Religious Affiliations and views	Memb. Freemasonry	Travels/Exile	Memb. In Scientific/Literary Soc.	Social Relations/Kinship
Cunha, Luís da (1662-1749) Portuguese	• Diplomat • Advocate of political, economic, cultural reforms based on Enlightenment and Mercantilism • Endowed the Royal Library and the Univ of Coimbra with new books	• Univ. Coimbra	• Catholic • Opposed clergy's influence and the Inquisition	--------	G. B. France Holland Spain Brazil	• Re-launched the Academy of the Generous (1685-1693)	Alexandre de Gusmão; King João V; Count of Ericeira Marquis of Pombal Alvares da Cunha (father)
Azevedo Fortes (1660-1749) Portuguese (French ancestry)	• Army Officer • Engineer • Populariser of Descartes and Locke in Port. • Defined the status of Engineering	• Universities in Spain and France	• Catholic	--------	Spain France Italy	• Acad. of the Discreet, Lisbon,(1696-1717) • Portuguese Academy, (1716-1717) • Royal Acad. of History, Lisbon, 1720	Bluteau; Count of Ericeira; Manuel de Campos; Alvares da Cunha; Luís da Cunha; M. Serrão Pimentel; L. Caetano de Lima; M. Caetano de Sousa
Count of Ericeira (1673-1743) Portuguese	• Polymath • Promoter of Science and founder of Scientific and Literary Academies	• Domestic education	• Catholic	--------	--------	• Acad. of the Generous, Lisbon • Acad. of the Discreet, Lisbon, (1696-1717) • Royal Acad. of History Lisbon, 1720	Bluteau; Azevedo Fortes Alvares da Cunha;Luís da Cunha; M. Serrão Pimentel L. Caetano de Lima; M. Caetano de Sousa

EIGHTEENTH-CENTURY PORTUGAL AND THE NEW SCIENCES

Name/ Nationality	Profession/ Contributions	Educational Background	Religious Affiliations and views	Memb. Freemasonry	Travels/ Exile	Memb. in Scientific/ Literary Soc.	Social Relations/Kinship
Bluteau, Rafael (1683-1734) French (born in G.B.)	• Oratorian Priest • Polymath • Promoter of Encyclopaedism	• La Flèche, Fra. • Jesuit Coll. of Clermont, Fra. • Univ. Verona • Univ. Rome	• Catholic	?	Italy, G.B., France Portugal: (1656-1697) (1704-1734)	• Acad. of the Discreet, Lisbon (1696-1717) • Acad. of Learned Conferences, Lisbon • Royal Acad. of History, Lisbon, 1720	Count of Ericeira; Boileau; Fontenelle; King João V; Azevedo Fortes
Castro Sarmento (1691-1762) Portuguese	• Physician • Transl. of Newton	• Jesuit Coll. Arts Évora, Portugal • Univ. Coimbra • Univ. Aberdeen	• Converso • Jew (persec. Inquisition)	--------	G. B. (exile in London, from 1730)	Royal Society	Count of Ericeira; M. A. Azevedo Coutinho (Amb. in London); Marq. of Pombal; Sachetti Barbosa; Desaguliers
Ribeiro Sanches (1699-1783) Portuguese (later French?)	• Physician/ Pedagogue • Inspired educational reforms	• Univ. Coimbra • Univ. Salamanca, Spain • Univ. Leiden, Holland	• Converso • Jew (persec. Inquisition) • Sympathetic to Protestantism	yes	Spain, France Holland, Russia, Turkey German States	• Acad. Sci. Paris • Acad. Sci. St. Petersburg • Royal Society • Acad. Sci. Lisbon	Boerhaave; Diderot; D'Alembert; Marquis of Pombal; Catherine II (Russia) Castro Sarmento; Chevalier Teodoro de Almeida; Brotero

Name/ Nationality	Profession/ Contributions	Educational Background	Religious Affiliations and views	Memb. Freemasonry	Travels/ Exile	Memb. in Scientific/ Literary Soc.	Social Relations/Kinship
Marquis of Pombal (1699-1782) Portuguese	• Diplomat • Prime-Minister to King José I (Absolut.) Implemented: . Political and economic reforms based on science and technol. . Educational reforms created the Coll. of Nobles; reformed the Univ. Coimbra . Created the Royal Printing Office	• Univ. Coimbra	• Catholic	• Initiated in G.B. (memb. in Port. Fremas. unclear)	G.B. Austria	• Lusitanian Arcadia	Count of Ericeira; King João V; Luís da Cunha; King José I; Ribeiro Sanches; Cenáculo; Verney; Dalla-Bella; Vandelli; Anastácio da Cunha; Monteiro da Rocha
Verney, Luís António (1713-1792) Portuguese (French ancestry)	• Clergyman • Theologian • Pedagogue • Disseminator of modern Philosophy (Bacon, Descartes, Newton, Locke)	• Coll. of Santo Antão, Lisbon • Univ. Evora • Univ. Rome	• Catholic	------	Brazil, Italy	• Acad. Sci. Lisbon	Giovanni Carbone; Muratori; Marquis of Pombal; Cenáculo; Genovesi; Facciolati; Duke of Lafões; J. J. Magalhães; Correia da Serra; / Chevalier (uncle)

Name/ Nationality	Profession/ Contributions	Educational Background	Religious Affiliations and views	Memb. Freemasonry	Travels/ Exile	Memb. in Scientific/ Literary Soc.	Social Relations/Kinship
Sachetti Barbosa (1714-1774) Portuguese	• Physician • Co-author of the statutes of the reformed Univ. of Coimbra, 1772 • Reformer of the Fac. Medicine, Univ. of Coimbra, 1772 • Founder of Medical Societies	• Univ. of Évora • Univ. of Coimbra	• Cath.(?)	------	------	• Medical Acad. of Madrid • Royal Society • Medical Acad. of Oporto, 1749	Castro Sarmento; Ribeiro Sanches; Emanuel Mendes da Costa (Clerk e librarian RSL); M. Gomes de Lima; Cenáculo
Duke of Lafões (1719-1806) Portuguese	• Army officer • Politician • Polymath • Encyclopaedist • Founder of the Royal Acad. of Scie., Lisbon, 1779	• Univ. of Coimbra	• Catholic	yes	Switzerland, France, Italy, Greece, Prussia, Poland, Eastern countries, Scandinavia	• Royal Society • Royal Acad. Sciences, Lisbon	Correia da Serra; Verney; various European intellectuals / • Queen Maria I, Port., (niece) • José Bonifácio (kin)
Magalhães (or Magellan), João Jacinto (1722-1790) Portuguese	• Augustinian Abbot • Physicist, Chemist • Instrument designer • Disseminator of scientific knowledge in Europe	• With the Augustinians, Coimbra	• Catholic	?	France, G. B., Austria, The Netherlands	• Acad. St. Petersburg • Acad. Sci., Brussels • Phil. Soc., Philadelphia • Acad. of Berlin ; Royal Society; Acad. Sci. , Paris; Chapt.Coffee House Soc. , London	Ribeiro Sanches; Volta, Lavoisier; Franklin; Euler; Priestley; Joseph Banks; William Hunter; W. Watson; J. Linch; Trudaine de Montigny; G. de Bory; R. Kirwan; Adair Crawford

Name/ Nationality	Profession/ Contributions	Educational Background	Religious Affiliations and views	Memb. Freemasonry	Travels/ Exile	Memb. in Scientific/ Literary Soc.	Social Relations/Kinship
Teodoro de Almeida (1722-1804) Portuguese	• Oratorian Priest • Teacher of Nat. Philosophy • Disseminator of modern science	• With the Oratorians (Phil. e Science)	• Catholic	------	Spain France	• Royal Acad. Scien. Lisbon (founder) • Real Soc. Vascongada Spain	Ribeiro Sanches; Correia da Serra; Duke of Lafões Verney; B. M. Portugal; Chevalier; M. A Vila.
Cenáculo, Friar Manuel do (1724-1814) Portuguese	• Franciscan • Pedagogue • Reformer: . Higher education (emphasis on experiment and Maths.) . Founder of various libraries . Implemented the instruction of teachers for girls in Beja . 1st. Museum Natural History. at Beja . President of the Royal Censorship Committee	• With the Oratorians • Univ. Coimbra	• Catholic	------	Italy	• Acad. Sci. Lisboa	Verney; Ribeiro Sanches; Monteiro da Rocha; Sachetti Barbosa; Ribeiro Sanches; Marquis of Pombal
Dalla Bella, Giovanni Antonio (1726-c.1823) Italian	• Physicist • Prof. Coll. of Nobles Lisbon • Prof. Univ. Coimbra	Univ. Padua	• Catholic? • Freethinker?	?	Italy Portugal	• Royal Acad. Sci. Lisbon (found. meb.) ?	Vandelli; Correia da Serra; Duke of Lafões

EIGHTEENTH-CENTURY PORTUGAL AND THE NEW SCIENCES

Name/ Nationality	Profession/ Contributions	Educational Background	Religious Affiliations and views	Memb. Freemasonry	Travels/ Exile	Memb. in Scientific/ Literary Soc.	Social Relations/Kinship
Vandelli, Domenico (1730-1816) Italian	. Prof. Univ. Coimbra . Industrialist . Botanist, Chemist . Established the Lab. of Chemistry, and the Botanical Garden, Univ. Coimbra	. Univ. Padua	. Catholic? . Freethinker?	?	Italy Portugal G. B. (exile 1810-1813)	. Royal Acad. Sci. Lisbon (found. memb.)	Dalla-Bella; Sobral; Seabra; Linnaeus; Correia da Serra; Duke of Lafões
Cunha, José Anastácio da (1744-1787) Portuguese	. Army-officer . Prof. Univ. Coimbra . Prof. Casa Pia, Lisb. . Mathematician: . Foundations of Mathemat. Analys.: original def. of a convergent series; new theory of the expon. function. . Poet. . Translator of Otway, Shakespeare, Gessner	. With the Oratorians, Lisb.	. Catholic . Freethinker (Persec. Inquisition)	yes	(contacts with Protestant foreign army-officers in the context of the reform of the Portuguese army carried out by the Prussian Count of Lippe)	----------	J. M. de Abreu; A. J. Rodrigues; D. A. Sousa Coutinho, Count of Funchal; M. P. Melo Rivalry with Monteiro da Rocha

Name/Nationality	Profession/Contributions	Educational Background	Religious Affiliation and views	Memb. Freemasonry	Travels/Exile	Memb. in Scientific/Literary Soc.	Social Relations/Kinship
Monteiro da Rocha (1734-1819) Portuguese	• Jesuit • Secular Priest • Co-author of the new statutes of the Univ. Coimbra • Prof. of Maths. and Astronomy Univ. Coimbra Astronomer: determination of the orbits of comets	• With the Jesuits in Brazil • Univ. Coimbra Canon-Law and Philosophy	• Catholic	------	Brazil	• Royal Acad. Sci., Lisb • Acad. Royal Navy, Lisbon	João Jacinto de Magalhães; Cenáculo; Marquis of Pombal
Brotero, Félix Avellar (1744-1828) Portuguese	• Clergyman • Prof. of Botany, Univ. Coimbra • Botanist: introduction of modern botanical ideas in Portugal; creation of a Portuguese botanical vocabulary; inventory of Port. flora	• With the Franciscans • Univ. Coimbra (law) • With Valmont de Bomare and Buisson, Paris (Botany) • Univ. Reims (Medicine)	• Catholic (Persec. Inquisition?)	yes	France,(exile) Holland, G. B. German States Italy	• Horticultural Soc., London • Linnean Soc., London • Soc. Philom., Paris • Soc. Hist. Nat. Paris • Physiog. Soc., Lunden • Nat. Hist. Soc., Rostock • Cesarea Acad., Bonn • Royal Acad. Sci. Lisb	Filinto Elisio; Vandelli; V. Sousa Coutinho (Amb. Paris); Ribeiro Sanches; Vicq D'Azyr; Daubenton; A. L. Jussieu; Buffon; Condorcet; Lamarck; Correia da Serra; H. Link; De Candolle; Sprengel; Hackel

Name/ Nationality	Profession/ Contributions	Educational Background	Religious Affiliations and views	Memb. Freemasonry	Travels/ Exile	Memb. in Scientific/ Literary Soc.	Social Relations/Kinship
Correia da Serra (1750-1823) Portuguese	• Clergyman • Botanist (trans-formist) • Politician • Founder of the Royal Acad. Sci., Lisbon • First Ambassador of Portugal in the USA (1816-1821)	• With Genovesi, Naples and in Rome	• Catholic	yes	Italy, G. B., France USA	• American Phil. Soc. • Acad.Belles Lettres • Soc. Philomatique, Paris • Acads. of Turin, Florence, Sienna • Acads. of Lyon, Bordeaux, Marseilles • Acad. of Liège • Royal Acad. Sci., Lisb • Wistar Party	Verney; Duke of Lafões; Joseph Banks; De Candolle; C. Gärtner; A. von Humboldt; Link; Brotero; Thomas Jefferson Genovesi
March. of Alorna (1750-1839) Portuguese	• Poet (scientific poetry) • Translator of Gray Wieland, Goethe, Young, Thomson	• Domestic educ. • Self-taught (tutored by Filinto Elísio)	• Catholic	? (connections with Freemasons)	Austria, London (exile) Spain France	• Lusitanian Arcadia	Filinto Elísio; Brotero; Teodoro de Almeida; King Carlos III of Spain; Necker; Madame de Staël Count of Oeynhausen, German Diplomat (husband)

Name/ Nationality	Profession/ Contributions	Educational Background	Religious Affiliations and views	Memb. Freemasonry	Travels/ Exile	Memb. in Scientific/ Literary Soc.	Social Relations/Kinship
Sobral, Tomé Rodrigues (1759-1829) Portuguese	• Clergyman • Prof. Chem., Univ. Coimbra • Chemist: adopted Lavoisier's chem.; translated Guyton de Morveau (affinity); applications of chem. to hygiene e defence	• Univ. Coimbra	• Catholic	?	-------	• Royal Acad. Sci. Lisbon	Seabra, Vandelli; Bonifácio ?
Bonifácio, José (1763-1838) Portuguese (Brazilian born)	• Prof. Metallurgy. Univ. Coimbra • Administ. of Mines. Portugal • Politician (independence of Brazil; abolitionist) • Mineralogist: identification of unknown minerals	• With Fr.M. Ressurreição, S. Paulo, Brazil • Univ. Coimbra • In Paris • In Freyberg	• Catholic • Freethinker	yes	France Italy German States Scandinavia Belgium England	• Soc. Philoma., Paris • Min. Soc. Iena • Acad. Berlin • Acad. Stockholm • Geol. Soc., London • Werner. Soc., London • Acad. Sci., Paris • Royal Acad. Sci., Lisbon	Haüy; Jussieu; Chapal; Fourcroy; Duhamel; Lempe; Köhler, Klotzsch, Freisleben; Lampadius; A. von Humboldt; Esmark; Del Rio; Volta Duke of Lafões (kin)
Seabra, Vicente Coelho (1764-1804) Portuguese (Brazilian born)	• Prof. Chem., Univ. Coimbra • Introduced Lavoisier's chem. in Portugal (1778); translated into Portuguese the new chemical nomenclat.	• Univ.Coimbra	• Catholic?	?	Brazil	• Roy. Acad. Sci. Lisb.	Sobral; Vandelli; Bonifácio

NOTES

[1] David Francis, *Portugal, 1715-1808* (London, 1985); David Goodman, "Portugal," in J. Yolton, et al. eds., *The Blackwell Companion to the Enlightenment* (Oxford, 1991), 420-421. See also entries "Pombal, Sebastião José de Carvalho," 412, "João V", 252, "José I", 252; Jeremy Black, "Lisbon", in J. Yolton, ed., *The Blackwell Companion to the Enlightenment* (Oxford, 1991), 291-292.

[2] João da Mota e Silva, known as Cardinal da Mota (1685-1747), was one of the most trusted advisers to King João V. He completed a doctorate in Theology at the University of Coimbra. Due to his solid humanist background he was entrusted by the King with the cataloguing of the books on theology that the monarch had acquired abroad. In 1736, he was officially appointed adviser to the king.

[3] The role of the aristocracy in the administration was greatly reduced during the reign of King João V, especially in Central and Southern Portugal. The reasons were the availability of great quantities of Brazilian gold, and of qualified civil servants outside the nobility. These allowed the monarch to implement strong disciplinary measures which limited the power of the aristocracy. However, the nobles of Northern Portugal and those holding overseas appointments were able to keep their autonomy even while continuing to challenge the King's power.

[4] Relations with the Papal Court deteriorated the monopoly of trade of the Orient, which troubled the international prestige of Portugal. The lobbying of Spain and France at the Papal Court reinforced these disagreements, and led to suspend the relations with Portugal. In 1732, this conflict was resolved, and diplomatic relations between both parts were re-established in 1737.

[5] J. Borges de Macedo, *O Marquês de Pombal (1699-1782)* (Lisboa, 1982); L. Azevedo, *O Marquês de Pombal e a sua Época* (Lisboa/Porto, 1909).

[6] The Jesuits arrived in Portugal in 1540 with the purpose of evangelizing the Portuguese territories in India. Their activities overseas were so highly praised that they finally established themselves in the Portuguese mainland. From the mid-sixteenth-century onwards, they opened various schools for the Christian education of youth. In effect they had the monopoly of the teaching system until the intervention of King José I, in 1759.

[7] The Oratorians, or Congregation of St. Philip of Neri, arrived in Portugal in 1668 with the support of Father Bartolomeu de Quental.

[8] Queen Maria I retired from state affairs in 1792, due to insanity. Her son became the regent until his mother's death, and was crowned as King João VI in 1816.

[9] England was in commercial difficulties due to the loss of its North-American colonies, and France and Holland saw their maritime power decreased after the French Revolution. In this way Portugal was able to explore the Cape Route.

[10] F. A. Oliveira Martins, *Pina Manique, o Politico - O Amigo de Lisboa* (Lisboa, 1948).

[11] J. H. Saraiva, *Discurso da Sessão Inaugural das Comemorações do Bicentenário da Casa Pia de Lisboa* (Lisboa, 1980).

[12] Roy Porter, Mikulas Teich, eds., *The Scientific Revolution in National Context* (Cambridge, 1992); Lorraine Daston, "The several contexts of the Scientific Revolution," *Minerva*, 32 (1994), 108-114; Dena Goodman, *The Republic of Letters: A cultural History of the French Enlightenment* (New York, 1994); Peter Jones, ed., *Philosophy and science in the Scottish Enlightenment* (Edinburgh, 1988); Henry E. Lowood, *Patriotism, profit, and*

the promotion of science in the German Enlightenment: The economic and scientific societies, 1760-1815 (New York, 1991); Jan Golinski, *Science as public culture: Chemistry and Enlightenment in Britain 1760-1820* (Cambridge, 1992); Joaquin Fernández Perez, Ignacio Gonzalez Tascon, eds., *Ciencia, tecnica y estado en la España ilustrada* (Madrid, 1990); A. Lafuente, A. Elena, M.L. Ortega, eds., *Mundializacion de la ciencia y cultura nacional* (Madrid, n/d); Antonio Lafuente, "L'organization de la science espagnole à l'époque des Lumières," in Xavier Polanco, ed., *Naissance et dévelopement de la science-monde* (Paris, 1989); Victor Navarro Brotons, "The Reception of Copernicus in Sixteenth-Century Spain: The case of Diego de Zuniga," *ISIS*, 86 (1995), 52-78; Anthony Padgen, "The reception of the "new philosophy" in 18th century Spain," *Journal of the Warburg and Courtauld Institutes*, 51 (1988), 126-140.

[13] David Goodman, "The Scientific Revolution in Spain and Portugal," in Roy Porter, Mikulas Teich, eds., *The Scientific Revolution in National Context* (Cambridge, 1992), 158-177.

[14] A.M. Amorim da Costa, "Chemical practice and theory in Portugal in the 18th century," in William R. Shea, ed., *Revolutions in science: their meaning and relevance* (Massachussets, 1988); A. M. Amorim da Costa, *Primórdios da Ciência Química em Portugal* (Lisboa, 1984); José Silvestre Ribeiro, *Historia dos Estabelecimentos Scientíficos, Litterários e Artísticos de Portugal nos Successivos Reinados da Monarquia*, 4 vols (Lisboa, 1871-1874).

[15] Rómulo de Carvalho, *História da Fundação do Real Colégio dos Nobres de Lisboa* (Coimbra, 1959), Rómulo de Carvalho, *História do Gabinete de Física da Universidade de Coimbra (1772-1790)* (Coimbra, 1978), Rómulo de Carvalho, *Relações entre Portugal e a Rússia no século XVIII* (Lisboa, 1979), Rómulo de Carvalho, *A Actividade pedagógica da Academia das Ciências de Lisboa nos séculos XVIII e XIX* (Lisboa, 1981), Rómulo de Carvalho, *A Física Experimental em Portugal no século XVIII* (Lisboa, 1982), Rómulo de Carvalho, *A Astronomia em Portugal no século XVIII* (Lisboa, 1985), Rómulo de Carvalho, *A História Natural em Portugal no século XVIII* (Lisboa, 1987), Rómulo de Carvalho, *História do Ensino em Portugal, desde a fundação da nacionalidade até o fim do regime de Salazar-Caetano* (Lisboa, 1986), Rómulo de Carvalho, *A Actividade Científica em Portugal no século XVIII* (Évora, 1996), Rómulo de Carvalho, "A pretensa descoberta da lei das acções magnéticas por Dalla-Bella em 1781 na Universidade de Coimbra," *Revista Filosófica*, 11 (1954), 103-138, Rómulo de Carvalho, "Portugal nas 'Philosophical Transactions' nos séculos XVII e XVIII," *Revista Filosófica*, 15-16 (1956), Rómulo de Carvalho, "Joaquim José Soares dos Reis, construtor das máquinas de física do Museu Pombalino da Universidade de Coimbra," *Vértice* 177 (1958), Rómulo de Carvalho, "Relações Científicas do Astrónomo francês Joseph-Nicolas de Lisle com Portugal," *Arquivo de Bibliografia Portuguesa*, 37-38 (1964-1966), 27-48, or Coimbra (1967), Rómulo de Carvalho, "A Física na Reforma Pombalina" in *História e Desenvolvimento da Ciência em Portugal-I Colóquio-até ao século XX* (Lisboa, 1986), Rómulo de Carvalho, "João Chevalier, astrónomo português do século XVIII," *Memórias da Academia das Ciências*, 32 (1992-93), 297-332, Rómulo de Carvalho, "Bento de Moura Portugal, homem de ciência no século XVIII," *Memórias da Academia das Ciências*, 33 (1993-94), 153-224.

[16] David Spadafora, *The idea of progress in 18th century Britain* (New Haven, 1990); K. Kumar, *Prophecy and Progress* (London, 1991).

[17] Luís de Albuquerque, Introduction to the translation of Copernicus, *As Revoluções das Orbes Celestes* (Lisboa, 1984).

[18] João Baptista, *Philosophia Aristotelica Restituta et illustrata quà experimentis quà ratiociniis nuper inventis* (Lisboa, 1748), 2 vols..

[19] Philiarco Phereponio, *Mercurio Philosophico* (Lisboa, 1752), 9.

[20] See Borges de Macedo, *Estrangeirados - um conceito a rever* (Lisboa, n/d);

[21] L. F. de Almeida, *A Propósito do Testamento Politico de D. Luis da Cunha* (Coimbra, 1948).

[22] O. M. Caldas, *D. Francisco Xavier de Menezes, 4º Conde de Ericeira. O Homem e a sua Época*, Ph. D. dissertation (Coimbra, 1958).

[23] Jean Bodin, *Methodus ad Facilem Historiarium Cognitionem* (Paris, 1566); Joseph Glanvill, *Plus Ultra, or the Progress and Advancement of Knowledge since the days of Aristotle* (London, 1668).

[24] Rafael Bluteau, *Prosas Portuguesas* (Lisboa, 1728), 39-40.

[25] Luís da Cunha played also a very important role in the acquisition of manuscripts, books, and atlases for the Royal Library and for the University of Coimbra.

[26] A. Gonçalves Rodrigues, "A correspondência científica de Dr. Sachetti Barbosa com Emmanuel Mendes da Costa, Secretário da Sociedade Real de Londres", *Biblos*, 14 (1938), 346-408; J. P. Sousa Dias, *A Água de Inglaterra no Portugal das Luzes*, M.Sc. dissertation (Lisboa, 1986).

[27] H. Cidade, *Lições de Cultura e Literatura Portuguesa* (6th ed., Coimbra, 1975), 2 vols.

[28] Rafael Bluteau, *Vocabulário de Lingua Portuguesa* (Lisboa, 1728), 2 vols.

[29] A. A. Banha de Andrade, "Manuel de Azevedo Fortes, primeiro sequaz, por escrito, das teses fundamentais cartesianas em Portugal", in *Contributos para a história da mentalidade pedagógica portuguesa* (Lisboa, 1982), 119-126; A. A. Coxito, *O Compêndio de lógica de M. Azevedo Fortes e as suas fontes doutrinais* (Coimbra, 1981).

[30] Manuel de Azevedo Fortes, *Lógica Racional, Geométrica e Analítica* (Lisboa, 1744).

[31] Jacob de Castro Sarmento, *Theorica Verdadeira Das Mares conforme à Philosophia do incomparável cavalhero Isaac Newton* (London, 1737).

[32] António Salgado Júnior (ed.), Luís António Verney, *O Verdadeiro Método de Estudar* (Lisboa, 1949-1952), 5 vols..

[33] The first practical application of the steam-engine to industry in Portugal occurred only in 1819. Preface by Jorge Custódio to the new edition of José Acúrcio das Neves, *Memória sobre os meios de melhorar a indústria portuguesa considerada nos seus diferentes ramos* (Lisboa, 1980), 67-68.

[34] Manuel de Azevedo Fortes, *O Engenheiro Portuguez* (Lisboa, 1728).

[35] Ribeiro Sanches, *Cartas sobre a Educação da Mocidade* (Paris, 1760). Although the front page mentions Cologne as the place of publication the book was, in fact, published in Paris.

[36] Ribeiro Sanches, *Método para Aprender a Estudar Medicina* (n/l, 1763).

[37] Francisco Gama Caeiro, *Frei Manuel do Cenáculo. Aspectos da sua Actuação Filosófica* (Lisboa, 1959); Jacques Marcadé, "D. Fr. Manuel do Cenáculo Vilas-Boas, provincial des réguliers du Tiers Ordre Franciscain", in *Arquivos do Centro Cultural Português* (Paris, 1971), vol.3, 431-458.

[38] Frei Manuel do Cenáculo, *Conclusiones de Lógicae* (Lisboa, 1751). Another edition with the text established, translated and annotated by João Pereira Gomes was published in Lisboa in 1958.

[39] J. Brücker, *Historia critica philosophiae mundi incunabulis ad nostram usque aetatem deducta* (Leipzig, 1741-1744) and *Institutiones historiae philosophiae* (Leipzig, 1747).

[40] J. Pereira Gomes, *Os Começos da Historiografia Filosófica em Portugal* (Lisboa, 1956), 11-15.

[41] Frei Manuel do Cenáculo, *Cuidados Literários* (Lisboa, 1791).

[42] Rómulo de Carvalho, *Colégio dos Nobres* (ref. 15), 189-190.

[43] J. G. Ferreira, *O Marquês de Pombal e as Reformas do Ensino* (Coimbra, 1989).

[44] Amorim da Costa, *Primórdios* (ref. 14); Giovanni Constanzo, "Fisici italiani in Portogallo", in *Relazioni Storichi fra l'Italia e il Portogallo* (Roma, 1940).

[45] Amorim da Costa, *Primórdios* (ref. 14).

[46] Portuguese currency at the time.

[47] The reputation of the Portuguese in Rome had been built on various episodes. During the reign of King Manuel I (1469-1521), extravagant embassies were sent to the Pope, displaying exotic plants and animals brought from the Portuguese colonies, inspiring several Italian artists. Later, King João V (1689-1750) spent large amounts of his Brazilian gold in order to obtain from Benedict IV the title of *Fidelisssimo*. The Pope's endorsement was an important piece of his external policy.

[48] J. A. S. Carvalho, *Memoria Historica da Faculdade de Philosophia* (Coimbra, 1872), 234.

[49] H. F. Link, *Voyage en Portugal depuis 1797 jusqu'en 1799* (3 vols., Paris, 1803-1805), 1, 300-301. About Heinrich Friedrich Link, see C. C. Gillispie, ed., *Dictionary of Scientific Biography* (New York, 1981), 7-8, 373-374.

[50] Dalla Bella, *Physices Elementa* (Coimbra, 1789-1790), 3 vols.

[51] J. Anastácio da Cunha, *Principios Mathematicos* (Lisboa, 1790).

[52] J. F. Queiró, "J. A. da Cunha: a Forgotten Forerunner", *The Mathematical Intelligencer* (1988), 38-43; A.J. Franco de Oliveira, "A. da Cunha and the Concept of Converging Series," *Archives for the History of Exact Sciences* (1988), 1-12.

[53] In 1811, Cunha's *Principios Mathematicos* was published in Bordeaux, translated by João Manuel de Abreu, a former pupil and friend: Joseph-Anastase da Cunha, *Principes Mathématiques (traduits littéralement du Portugais par J.M. Abreu)* (Bordeaux, 1811). The same book was re-published in Paris in 1816: Joseph-Anastase da Cunha, *Principes de Mathématiques (Savoir: ceux de l'Arithmétique, de la Géométrie, de l'Algèbre, de son Application a la Géométrie et du Calcul Différentiel et Intégral, exposés d'une manière entièrement nouvelle)* (traduits littéralement du Portugais par J.M. Abreu) (Paris, 1816).

[54] A. P. Youskevitch, "J. A. da Cunha et les Fondements de l'Analyse Infinitésimale," *Rev. Hist. Sci.*, 26 (1973), 3-22 and "C. F. Gauss et J. A. da Cunha," *Rev. Hist. Sci.*, 31 (1978), 327-332.

[55] Júlio Henriques, "Felix Avellar Brotero," *O Instituto*, 37 (1889), 367; A. Fernandes, "Felix Avellar Brotero e a sua obra," *Boletim da Sociedade Broteriana*, [2], 19 (1944), 53-76.

[56] Felix Avellar Brotero, *Compêndio de Botânica* (Paris, 1788).

[57] Felix Avellar Brotero, *Principios de Agricultura Philosophica* (Coimbra, 1793).

[58] Felix Avellar Brotero, *Flora Lusitanica* (Lisboa, 1804).

[59] Felix Avellar Brotero, *Phytographia Lusitaniae Selectior* (Lisboa, 1816, vol. 1; 1827, vol. 2).

[60] Carlos A.L. Filgueiras, "Vicente de Seabra Telles (1764-1804), the first Brazilian chemist," *NTM*, 27 (1990), 27-44; Amorim da Costa, "Chemical practice" (ref. 14); Amorim da Costa, *Primórdios* (ref. 14).

[61] Vicente Coelho de Seabra, *Elementos de Chimica* (Coimbra, 1788-1790).

[62] Scopoli, *Fundamenta Chemiae* (Prague, 1777).

[63] Vicente Coelho de Seabra, *Nomenclatura Chimica Portugueza, Franceza e Latina a que se junta o sistema de characteres chimicos adaptados a esta nomenclatura por Hassenfratz e Adet* (Lisboa, 1801).

[64] Edgard de Cerqueira Falcão (ed.), *Obras Científicas, Políticas e Sociais de José Bonifácio de Andrada e Silva* (S. Paulo/Brazil, 1965), 3 vols.; Octávio T. de Sousa, *José Bonifácio* (S. Paulo/Brazil, 1965). His full name was José Bonifácio de Andrada e Silva, but he is best known as José Bonifácio.

[65] The four new species of minerals were respectively called: Petalite, Spodumene, Criolite, and Scapolite. The eight varieties he thought of as new species were: Wernerite (a variety of scapolite), Acanthicone (a variety of epidote), Salite (a name which applies today to common pyroxenes), Coccolite (a variety of pyroxenes), Ichtyophtalme (a mixture of apophyllite with indicolite and allochroite), Indicolite (blue tourmaline), Aphrizite (a variety of black tourmaline) and Allochroite (garnet).

[66] José Bonifácio (Herrn d'Andrada), "Kurze Ungabe der Eigenschaften und Kennzeichen einiger neuen Fossilien aus Schweden und Norwegen, nebst einigen chemischen Bemerkungen über dieselben," *Allgemeines Journal der Chemie*, 4 (1800), 29-39.

[67] José Bonifácio, "Short notice concerning the properties and external characters of some new fossils from Sweden and Norway; together with some Chemical remarks upon the same," *Journal of Natural Philosophy, Chemistry and the Arts* (1801), 193-193; "Exposé Succint des caractères et des propriétés de plusieurs nouveaux mineraux de Suède et de Norvège, avec quelques observations chimiques faites sur ces substances," *Journal de Physique, de Chimie, d'Histoire Naturelle et des Arts*, 51 (1800), 239-246.

[68] J. S. M. Leal Júnior, *Elogio Histórico de D. João Carlos de Bragança, Duque de Lafões* (Lisboa, 1859).

[69] J. Henriques, "José Francisco Correia da Serra," *Boletim da Sociedade Broteriana*, 1 (1922), 85-125; J. E. Agan, "Corrêa da Serra," *Boletim da Sociedade Broteriana*, [2], 4 (1926), 9-43.

[70] Augustin de Candolle, *Théorie élémentaire de botanique*, according to quotation in *Revista Universal Lisbonense* (1844), 565.

[71] *Philosophical Transactions of the Royal Society, Transactions of the Linnean Society, Transactions of the American Philosophical Society, Annales du Muséum, Bulletin de la Société Philomatique, Archives Littéraires de l'Europe* and *The American Review*.

[72] F. C. Domingues, *Ilustração e Catolicismo. Teodoro de Almeida* (Lisboa, 1994).

[73] Teodoro de Almeida, "Oração na abertura da Academia das Sciencias em 1 de Julho de 1780," in Christovam Ayres, *Para a História da Academia das Sciencias de Lisboa* (Coimbra, 1927).

[74] Francisco Contente Domingues, "Science and nationalism: Portugal in the late 18th century," *History of European Ideas*, 16 (1993), 91-96; Domingues, *Ilustração* (ref. 72).

[75] The group of mathematicians led by Cunha was formed by:
- João Manuel de Abreu (1757-1815), who met Cunha at the Artillery Regiment of Oporto. In 1784, he attended as a volunteer the first two years of the course on Mathematics at the University of Coimbra, where he completed his studies in 1787. Later, he became lecturer at the Academy of the Royal Navy and professor of History at the College of Nobles. He published in France, in 1809, *Supplément à la traduction de la Géometrie d'Euclide, de Mr. Peyard, publiée en 1804; et à la Géometrie de Mr. Legendre: Suivi d'un essai sur la vraie théorie des parallèles*, the publisher being A. Agen.
- Anastácio Joaquim Rodrigues (?-1818), an army officer who became disciple and friend of Cunha. He was considered as one of the most gifted Portuguese mathematicians of his time. He was lecturer at the Academia Real de Fortificação (Royal Academy of Fortification), and in 1814, he became member of the Academia Real das Ciências de Lisboa.
-Domingos António de Sousa Coutinho, 1st Count of Funchal (1760-1833) had a degree in Law from the University of Coimbra and held the post of Portuguese Ambassador in Lon-

don. He had initiated his diplomatic career as a Portuguese diplomatic representative in Copenhagen and then in Turin and Rome.
- Manuel Pedro de Melo (1765-1833) a pupil of Cunha at the Casa Pia. He then was admitted to the College of Natural Sciences in Coimbra. In 1789, he enroled in the University of Coimbra and, in 1795, he completed a degree in Mathematics. In 1798, he finished a degree in Medicine. In 1801, he became professor of Hydraulics at the University of Coimbra. He established friendly relationships with various European men of science. These contacts helped him in the acquisition of several scientific instruments, which were gathered together in the Physics Cabinet of the University. In 1806, he was awarded a prize for his memoir "Memória sobre o problema do paralelogramo de forças" ("Memoir on the question of the parallelogram of forces") by the Royal Academy of Copenhagen. He also contributed various articles to the *Annales des Arts et des Manufactures*, and to the publications of the Academia Real das Ciências de Lisboa. Melo's work was praised by Adrien Balbi, *Essais de statistique sur le Royaume de Portugal et d'Algarve, comparé aux autres états de l'Europe - Appendix à la Géographie Littéraire* (Paris, 1822).

[76] Isabel Malaquias, Manuel Fernandes Thomaz, "Scientific Communication in the 18th century: The case of John Hyacinth de Magellan," *Physis*, 31 (1994), 817-834; Stephen F. Mason, "Jean Hyacinthe de Magellan, F.R.S., and the chemical revolution of the 18th century," *Notes Rec. Roy. Soc. Lond.*, 45 (1991), 155-164; Stuart Pierson, "Magellan, Jean-Hyacinth," *Dictionary of Scientif Biography* 9 (1981), 5-6; Isabel Malaquias, *A Obra de João Jacinto de Magalhães no Contexto da Ciência do Séc. XVIII*, Ph. D. dissertation (Aveiro, 1994).

[77] J. J. Magalhães, *Essai sur la nouvelle théorie du feu élémentaire et de la chaleur des corps* (Paris/London, 1780).

[78] This second edition was intended for publication in 1780, but in fact was only printed in 1788. This publication greatly annoyed Kirwan, who in the meantime had managed to publish his own *Mineralogy* (1784).

[79] José Tengarrinha, *História da Imprensa Periódica Portuguesa* (2nd. ed., Lisboa, 1989); Maria de Fátima Nunes, *Leitura e Agricultura: a imprensa periódica científica em Portugal (1772-1852)*, Ph.D. dissertation (Évora, 1994), 2 vols..

[80] "Dedicatória a D. João de Almada e Melo," *Gazeta Literária*, 2 (1762), 10. João de Almada e Melo was the Mayor of Oporto.

[81] Marquesa de Alorna, *Recreações Botânicas* in *Obras Poéticas de D. Leonor de Almeida Portugal e Lencastre* (Lisboa, 1844), 6 vols. (vol. 4).

[82] Teodoro de Almeida, *Recriação Filosófica* (Lisboa, 1751-1800), 10 vols.

[83] Teodoro de Almeida, *Cartas Fysico-Mathematicas* (Lisboa, 1784-1798), 3 vols.

[84] Memoria sobre a Natureza da luz, e Vacuo Celeste, *Arquivo da Academia das Ciências*, ms. 377 azul nº 2; Memoria sobre a natureza do Sol, *Arquivo da Academia das Ciências*, ms. 377 azul nº 1; Memoria sobre a rotação da lua, *Arquivo da Academia das Ciências*, ms. 352 azul nº 4; Memoria sobre huma maquina para conhecer a causa Fysica das Marés, *Arquivo da Academia das Ciências*, ms. 377 azul nº 4.

[85] Domingues, *Ilustração* (ref. 72).

[86] Benito Feijoo, *Cartas eruditas, y curiosas, en que, por la mayor parte, se continuà el designio del Theatro Critico Universal* (Madrid, 1742-1745), 2 vols.

[87] Teodoro de Almeida, *O Feliz Independente do Mundo e da Fortuna ou Arte de Viver Contente em Quaisquer Trabalhos da Vida* (Lisboa, 1779).

[88] Teodoro de Almeida, *Recriação* (ref. 82), vol. 1, 46-47.

DIMITRIS DIALETIS, KOSTAS GAVROGLU, MANOLIS PATINIOTIS

THE SCIENCES IN THE GREEK SPEAKING REGIONS DURING THE 17TH AND 18TH CENTURIES

The process of appropriation and the dynamics of reception and resistance

INTRODUCTION

What has been known as the Scientific Revolution of the 16th, and especially the 17th century was an exclusively European phenomenon. While the social, ideological, conceptual, theological, economic, and political repercussions of the new ideas developed during the Scientific Revolution have been systematically studied within the setting of the countries where that revolution originated, only few historical works have dealt with the issues related to the introduction of these ideas to the countries in the periphery of Europe (that is, the countries of the Iberian Peninsula, the Balkans, the Eastern European and the Scandinavian countries). How did the ideas of the Scientific Revolution migrate to these countries? What were the particularities of their expression in each country? What were the specific forms of resistance to these new ideas and to what extent did they display national characteristics? What were the legitimising procedures for the acceptance of the new ways of dealing with nature? Did the discourse used by the scholars for writing and discussing scientific issues share the same features as the discourse developed by their colleagues in the countries of Western Europe? Any attempt to understand the assimilation of the ideas of the Scientific Revolution in these regions and to assess the characteristics of the resistance to such an assimilation—especially during the Enlightenment— cannot omit discussion of at least some of these questions.

In this paper we shall discuss some of the issues related to the introduction of the new sciences to the Greek speaking regions during the 18th century.[*] Necessarily, we will also have to refer to developments in the 17th and 19th centuries. Such a discussion calls for a contextual approach: it cannot be conducted independently of an

---

[*] All the scholars wrote and taught in Greek. The regions where they lived and worked were quite often of a rather complicated ethnic composition. In any case, we should stress that the regions we are referring to are by no menas the same as those forming part of modern Greece. In many instances Greek scholars lived and worked in societies where there was a dominant element of Greek speaking merchants, e.g. Vienna, Leipzig, Venice, etc.

overall historical assessment of what it means for ideas that originated in a particular cultural and historical setting to have been «transmitted» into a different cultural milieu with different intellectual traditions and different political and educational institutions.

We find the concept of the «transfer» of ideas to be ultimately inadequate in contextualizing the dissemination of the new sciences in the Greek speaking populations, and in this paper we hope to indicate that «appropriation» can be a more coherent and fruitful analytic instrument. Appropriation directs attention to the measures devised *within the appropriating culture* in order to shape the new ideas within the local traditions which form the framework of local constraints –political, ideological, as well as intellectual constraints.

A historiography based on the concept of transfer can easily degenerate into an algorithm for keeping tabs on what is and what is not [«successfully»] transmitted. A historiography built around the concept of appropriation is more comparable to the procedures of cultural history; acceptance or rejection, reception or opposition are intrinsically cultural processes. Such an approach also permits the newly introduced scientific ideas to be treated *not* as the sum total of discrete units of knowledge but as a network of interconnected concepts. The practical outcome of a historiography of appropriation is to be able to articulate the particularities of discourse that is developed and eventually adopted within the appropriating culture.

Undoubtedly the concept of transmission of ideas is of some use to the historian of ideas. This, however, is apparent only when the transmission of ideas is used for certain specific cases within a wider context of the *appropriation* of multiple cultural traditions by the Greek speaking societies during a specified period of their history. In these occasions the intellectual and institutional framework for the reception of the new ideas is, to a large extent, conditioned by the cultural and religious traditions of the Greek speaking societies together with the role and structure of their educational institutions.

The purpose of the present inquiry, then, is not to examine the success of a process of transfer, but to study the production of a distinctive scientific discourse through the reception of the Western ideas within the Greek speaking regions during the 18th century. Members of a national community which was under occupation and had no state institutions of its own were able to appropriate new scientific ideas in forms that could function within a specific cultural milieu. A synthesis of elements of ancient Greek thought with christian Orthodox tradition had already emerged by the 18th century as a strong cohesive element in the intellectual identity of the Greek nation; the legitimation of the new scientific ideas ran parallel with economic and political restructuring, both assisting in the formation of a new coherent ideology and political stand, connecting the past of the Greeks with their future prospects as independent nation.

The history of ideas of the Greek speaking regions in the Ottoman Empire from the fall of Constantinople in 1453 to the Greek Revolution of 1821 is invariably linked with the educational policies articulated by the Orthodox Church and the Ecumenical Patriarchate. Simply put, the sciences in the educational institutions — which were under the jurisdiction of the Church throughout this period — were introduced as part of a modern curriculum which also (re)introduced ancient Greek thought as the precursor of all the glorious developments in Europe. The introduction of the sciences served both to "enlighten" the youth as well as to help create a national consciousness through the establishment of an intriguing continuity: from the ancients through Byzantium to the present leading to a future when "glory" will be re-established again in Greece. Thus, from the early years of the 17th century, the introduction of the sciences was subservient to the political and, to a certain extent, ideological reorientations of the Church and of the newly emerging social groups.

In this paper we would like to argue the following points.

1. Most analyses of the Scientific Revolution and the establishment of the new sciences in the various countries in Europe take into consideration a host of questions related to the formation of state institutions. Issues, for example, concerning patronage, the establishment of academies and the usefulness of the new sciences for economic production are couched within the context of the formation of state institutions. The situation was radically different in the Greek speaking regions and the Balkans which were under the Ottoman domination. Quite naturally, apart from the Church, the Greeks did not have any centrally administered institutions. In the study of the introduction and reception of the sciences a series of complicated issues enter the picture, especially since the Ottoman administration had granted to the Greek Orthodox Church the responsibility for the education of the Christian population. The content, however, of what was taught was not solely determined by the Church. It was, rather, the confluence of largely similar but at times conflicting aims of the religious hierarchy, of the social groups with significant economic activity and of the scholars themselves. And in order to comprehend what appeared to be a unified educational policy of the Church, it becomes necessary to articulate the relatively autonomous agendas of each of these religious and social groups.

2. In introducing the new sciences, the Greek scholars did not attempt to introduce natural philosophy *per se*, but, rather, they sought a new way of philosophising. This discourse lacked the constitutive features of the discourse of natural philosophy as it was being articulated and legitimised in Western Europe: it was primarily a philosophical discourse. Even while writing about the recent scientific developments, the Greek scholars of the Enlightenment thought of themselves first and foremost as philosophers. They did acknowledge the uniqueness of the developments in Western Europe concerning the new sciences; but they did not perceive these developments as a break with the approach of the ancient Greek philosophers.

The new sciences were, on the whole, interpreted as an expected corroboration of the programmatic declarations of ancient Greek philosophers. In introducing the new scientific ideas, they were reluctant to adopt the discourse used by the natural philosophers in the academic centres of Western Europe. It is only within such an interpretative framework that one can comprehend the *absence* of any discussion concerning the character of the rules of the new ways to study nature, the processes of legitimising the new viewpoint, and the initiation of consensual activities to consolidate the new attitude about the ways of dealing with natural phenomena.

3. Most of the books on science written by Greek scholars in the period of the Enlightenment were intended for use in education, and it has been the case that they seldom present scientific theories with the rogor expected of pedagogical texts written in the same period in Europe. Such discrepancies with respect to European norms have often been attributed to uneven scientific competence, a «watering down» of science at the periphery. We argue that this interpretation is uncalled for; it arises as a more or less direct consequence of the explanatory concept of «transfer», which locates the agency on the side of the culture in which the ideas originated. If one adopts the notion of transfer, this implies a kind of filtering process: there is a selective procedure in the transferring since it depends on who transfers the ideas, when and for what purpose they are transmitted. The study, then, of the introduction of the sciences is often reduced to accounts of what is held by the filter. But, such a viewpoint undermines the specificity of the sciences in the Greek speaking regions, that is, the subtle transformations of the scientific ideas during their process of appropriation.

4. In the centre, the main role of scientists was to produce scientific knowledge whereas their role in the periphery — perhaps with the exception of the Scandinavian countries — was entirely different. It was to disseminate this knowledge through the educational structures. Thus the predominantly *productive* role of the scientists in the centre has to be contrasted with the predominantly *educational* role of the scholars in the periphery. The educational agenda of the scholars played a rather decisive role since the discussion and the dissemination of the sciences was being exclusively realised within the educational institutions and many a times in reference to issues pertaining to education.

THE YEARS AFTER THE FALL OF CONSTANTINOPLE.

The last century of the Byzantine Empire witnessed works in astronomy, mathematics, alchemy and, of course, philosophy by scholars who formed the intellectual elite of a society fraught with religious disputes and political struggles. The exodus of these scholars and their migration, mainly to Italy and France, had started quite some time before the fall of Constantinople in 1453. In most instances, these scholars found their new environments quite congenial and in most cases they adopted a rather sympathetic position towards Catholicism. In the decade preceding the Fall, the dominant political forces had come to favor between the Eastern and Western Churches — even though Rome would clearly have the upper hand — as holding out the only hope of rescuing Eastern Christendom from the Ottomans. It was a move aimed at convincing the Catholics — nearly two centuries after the catastrophic siege and occupation of Constantinople by the forces of the fourth crusade between 1204 and 1261 — to rally for the rescue of Eastern Christianity. In contrast, the scholars who remained in Constantinople and initiated the re-establishment of the educational institutions, were, as a rule, carriers of an anti-western attitude.

Immediately after the fall of Constantinople, the Sultan Mohammed II — Mohammed the Conqueror — allowed the Patriarchate to continue its function. The Patriarch was recognised by the Sultan as the legal head of the Orthodox *millet* (nation). Most importantly, the Patriarchate was granted full jurisdiction over the education of the Orthodox Christian populations in the Ottoman Empire and, eventually, the right to develop mechanisms to collect some form of taxation from the parishes, while at the same time it was set free of tax obligations towards the Ottoman state. The Sultan's decision was, partly at least, a response to the contingencies of the new era of the Ottoman Empire. The serious difficulties relating to the administration of a continuously expanding Empire with a progressively increasing Christian population and the threat from Christian Europe could be eased by granting this limited political autonomy to the Patriarchate as well as by taking advantage of the deep animosity between the Orthodox Church and Rome. The Orthodox Church was at the time the only organised institution which could represent Christian nations of the Balkan Peninsula in their dealings with the Ottoman administration. Furthermore, the Patriarchate had already created a structured ecclesiastical hierarchy which allowed control of even the smallest Christian community in the area. Moreover, the choice of the Patriarchate as the *de facto* political representative of the Christian populations of the Balkans would contribute to the consolidation of the Ottoman Empire's newly acquired lands in the West. During that period, the idea of a European crusade against the Ottoman Empire under the aegis of the Papacy was a strong element of political coherence in threatened Europe. By setting up the Patriarchate as an autonomous political institution, as well as strengthening the forces contrary to the union of the Churches, Mohammed intended to minimize the possibilities for the realisation of such plans.

The Sultan appointed as the new Patriarch Gennadios Scholarios (1400-1460). It should be noted that since 1450 the Patriarchate had been in effect headless, while the nominal Patriarch Gregory III was in Rome. Gennadios Scholarios was a well-known jurist, rhetor and philosopher, and played an important role in political life during the last years of the Byzantine era. As a philosopher, he was of aristotelian orientation, a follower of Aquinas and an opponent of Pletho's platonism. Among his writings we note the *Synopsis of Aristotle's Physics with Simplicio's Commentaries* and *Against Pletho's Queries of Aristotle*. He was an officer of the Byzantine state, and a member of the delegations that represented the Orthodox Church during the negotiations with the Catholics. Though at first an advocate of the union of the Churches, soon afterwards, he became a fierce opponent, attacking the attempts of the emperor Constantinos Paleologos to come to terms with Pope Nicholas V. Historians agree that the Emperor's motives in agreeing to the union of the Churches were based less on religious grounds and more on the hope of securing the military support of the western countries against the Ottoman threat. Especially during the siege of Constantinople, Gennadios Scholarios made an intense propaganda against the Catholic Church. When Mohammed conquered Constantinople he saw in Gennadios the person to become the first Patriarch. Gennadios undertook the task of reviving the intellectual life of the city. He founded the first official school, the Patriarchal Academy, which was the continuation of the Pandidakterion of the Byzantine era, and appointed Mathaeos Kamariotis as its first director.

The exact date of Kamariotis' birth is unknown; he remained director of the Patriarchal Academy up to his death in 1489 or 1490. There is no information concerning the curriculum of the Academy. However, it is reasonable to suppose that it was similar to the standard Byzantine curriculum, since Kamariotis had been a well-known teacher for many years before the fall. From the part of his work that is known to us today we can see that he had then included in the curriculum ancient Greek literature, rhetoric and theology, while he was an opponent of Pletho's ideas.

By the end of the 16th century and within the context of counter-reformation after the Council of Trent (1545-1563), Rome defined a new policy towards the Greek population of the Ottoman Empire, designed to prevent any rapprochement between the Protestants and the Orthodox. In the beginning of the seventeenth century, the Patriarchates of both Constantinople and Jerusalem became fields of contention between the Catholics and the Protestants. The Jesuits attempted to create Catholic zones in the eastern Mediterranean basin and to this purpose they cooperated with the Ottomans, attempting at the same time to bribe several officers of the Orthodox Church.

The College of Saint Athanassios (Collegio Greco), in Rome, where many Greek scholars — who later in their lives became quite eminent — attended classes, played an important role in the development of the political influence of the Catholics. This College was founded in 1577 and its main mission was to offer higher education to the Greeks of the Ottoman Empire through study and acquaintance of the world of the Catholic Church. Theology according to the teachings of Aristotle and Thomas Aquinas was the basis of the curriculum. The doctrines of the Catholic Church, ancient Greek and Latin literature, Aristotelian philosophy and advanced theological education were included in the curriculum. Graduates before returning back to their country of origin, were asked to propagandise Catholicism, to cultivate Greek letters, and to support anti-Ottoman ideas, since the Ottomans were still viewed as the main enemy of the Christian world.

During the same period, the Protestants were trying to increase their influence in the eastern Mediterranean, especially through the activity of the ambassadors of England, Holland, Germany, and Sweden. Not unexpectedly, they offered support to the Patriarchates of Constantinople and Jerusalem. Their shared hostility to Catholicism brought the Protestants and the Greek Orthodox close to each other. In 1620 Kyrillos Loukaris (1572-1638) became Patriarch of Constantinople. During the early stages of the 30-year war, Loukaris planned a series of political moves to consolidate the survival of the Orthodox Church. He felt that there were unmistakable signs of an impending alliance between Catholic France and the Ottomans. He saw such an alliance as the main danger against the Orthodox Church, and he sought supporters among the Protestants, especially the Dutch. The ambassador of Holland to the Ottoman Court turned out to be a very co-operative ally. Loukaris, also, proceeded to write a notorious leaflet arguing for the common theological grounds between Calvinism and Orthodoxy. Many serious theologians — and not only his adversaries — accused him of adopting Protestantism.

Being convinced that the Catholic propaganda was effective because of its educational institutions, Loukaris upgraded the Patriarchal Academy and introduced what came to be known as «religious humanism». He himself had studied at the Greek School of Venice, under Maximos Margunios, from 1584 until 1588 and he had completed his studies at the University of Padua in 1593.

Religious humanism was an attempt to synthesize the teachings of ancient Greeks with the teachings of the Orthodox Church fathers, considering the intellectual traditions originating in Greek antiquity and those of Christianity as a unity. Religious humanism became the means for moulding a kind of national consciousness by reclaiming hellenistic roots through Greek Orthodox Christian teaching. In the prevailing conditions of intense national reorientations and regroupings in Europe, such a strategy aimed at upgrading the political role of the Patriarchate by providing an institutional expression to the ties between orthodoxy and hellenism. Such initiatives led not only to the establishment of new educational institutions, but, eventually, to the furthering of the Church's dominance through the articulation of a new ideological and political agenda. The idea that the Orthodox Church must safeguard the great intellectual tradition of the nation and protect Hellenism from the "Ottoman despot *and* the propaganda or the contrivances of Catholicism" was given a theoretical justification and an institutional expression.

The strong and systematic reference to the ancients eventually created a rather restricted space for lay theology and this was to prove quite important for the way the Greek scholars would collectively decide to promote the new scientific ideas. Establishing new schools with new curricula had a very specific purpose: to keep alive and modernize a national culture whose constitutive domains were both the ancients and Orthodox Christianity. In this context, the new scientific ideas were, at least partly, introduced as a means of underlining the success of the ancients' ideas.

In 1622 Kyrillos Loukaris appointed a renowned neo-Aristotelian, Theophilos Korydalleas (1570-1646), to the directorship of the Patriarchal Academy. The latter had studied in Italy during the first decade of the seventeenth century. In 1604 he attended classes at the Greek College in Rome. He went on to study at the University of Padua, at a time when Cesare Cremonini was the dominant figure and the articulate defender of Aristotelianism, especially against the new science introduced by his colleague there, Galileo Galilei. Korydalleas received his doctorate in Philosophy and Medicine, around 1608. In the Patriarchal Academy Korydalleas reorganised teaching along the ways practised in Padua. A central place was assigned to philosophy — as distinct from theology — and to the interpretation of the commentaries on the main works of Aristotle. Korydalleas' humanistic brand of philosophy contained the potential for a rupture with a strictly theological approach to nature and to human affairs. But at the same time, there was a conscious policy to contain and develop this new approach *exclusively* within the framework of neo-Aristotelianism, during a period when such a framework was being undermined and redefined elsewhere in Europe.

Theophylos Korydalleas, even though he was well acquainted with Descartes' philosophical and scientific works and he had, undoubtedly, come into contact with the ideas and the debates about heliocentrism during his stay in Italy, promoted the Aristotelian worldview. Nowhere in his works did he mention the heliocentric system, not even in order to criticize it. In 1626 he is thought to have written his *Global Geography According to Ptolemy*. This work was never published but references to it from his students indicate that it was a piece which revived the interest of the Greek scholars in the work of the Alexandrian astronomer, fifteen centuries after its first presentation. Korydalleas, however, held a viewpoint different from Ptolemy's. While the latter sought a mathematical description and modelling of the celestial movements, Korydalleas focused on physical, qualitative description, pointing out the Aristotelian features of the Ptolemaic cosmography. The Earth was at the centre of the universe because, as Aristotle taught, heavy bodies tended to fall towards the centre. The celestial spheres moved around the earth, depicting the celestial hierarchy. He also insisted on the main Aristotelian separation between the eternal and perfect world beyond the Moon and the world of change below it. At the end of the text, Korydalleas provided a brief descriptive account of the Ptolemaic theory of epicycles.

Korydalleas, through his teachings, contributed decisively to the gradual fusion between the Orthodox Christian theology and aristotelianism. The new methods of mathematical physics and the quantitatively oriented study of nature are nowhere to be found among his works. Against the unified, homogeneous, deterministic, and infinite universe of the newly articulated cosmology, he juxtaposed the finite and closed aristotelian world, as well as a geocentrism inspired by theology. His writings continued to exert a strong influence for a long time even though some of the later scholars were not eager to endorse all of his views, especially since he had also been accused of Calvinism and atheism. Nevertheless, Korydalleas was the first teacher after the fall of Constantinople who provided a frame of reference to be available to Greek educators in the coming years; and, despite the criticism directed against him, he continued to teach under the aegis of the Church.

INTRODUCING THE NEW SCIENTIFIC IDEAS

Most of the second half of the 17th century and a large part of the 18th century was a period of educational and economic rejuvenation of many sectors of Greek society. In this period the Fanariots — basically the Greeks who lived in Constantinople — would play a dominant role. The beginning of this period is characterized by the completion of the Ottoman expansion and the creation of some of the prerequisites for the economic development of a new Hellenism. From the end of the 17th century, the Fanariots acquired an increasingly important role in the administration of the Ottoman state. At the outset of the 18th century, representatives of the Fanariots were appointed by the *Sublime Porte* as governors and hospodars in Wallachia and Moldavia. The Fanariots would soon take the lead among all the other Greeks dispersed in the Balkans; their political dominance would reinforce the already strong influence of the Greeks in the economic as well as cultural spheres in these regions, while at the same time as administrators and as diplomats they would take the line commonly referred to as the enlightened despotism.

This period is characterised by three interdependent developments. First, the increasing involvement of this group of Greeks in the administrative affairs of the Ottoman Empire undermined the almost exclusive role of the clergy in mediating the relations of the Christians with the Court. The second characteristic of this period is the increasing receptivity to the new ideas coming from Europe by the Fanariots, whose relative autonomy from the Patriarchate was further strengthened by an agenda of «europeanization». The third characteristic is related to the rise of a new social group. In addition to the Fanariots, the merchants started to assert themselves socially and played a rather significant role in the intellectual orientations of the period. The symbiotic relationship between the merchants and the quasi-administrative group of Fanariots was not always without conflict. Often, for example, they were at odds concerning the exertion of influence on the Patriarchate. The social and economic prominence of these groups slowly led to the weakening of the absolute control the Church had on the schools and in their curricula. The Fanariots, for example, took many initiatives for the establishment of new schools.

At the same time, the Greek scholars started moving all over Europe. Italy ceased to be the almost exclusive place for their studies. Greek scholars started travelling to the Germanic countries, Holland, and Paris. They were, thus, intellectually influenced by a multitude of traditions and schools — and that was true for their training in the natural sciences as well. Interestingly, it was during that period that we witness a strong tendency of the scholars to return home after the completion of their studies abroad.

There were, basically, two reasons favouring the return of the scholars. The first was the growing need for scholars and teachers in the schools which were being founded as a result of the economically thriving Greek communities dispersed in various regions within the Ottoman Empire and outside it. Especially after the mid-1700s there was a upsurge in the establishment of new centres of "higher education" in the Greek communities in great mercantile centres like Venice or Vienna, in Wallachia and Moldavia, in cities containing sizeable Greek administrative communities, like Bucharest and Iasi, in the Ionian Islands and in cities with large Greek communities, like Ioannina, or cities on the coast of Asia Minor. Individuals or groups of people, very often expatriates, gave money or bought books and equipment for these schools. In some schools, like in Chios and Milies, these endowments resulted in remarkable libraries. These schools, as centres of intellectual activity and as expressions of educational and patriotic philanthropy became paradigmatic of the Greek Enlightenment.

The second reason for the return of the scholars had to do with their gradual marginalization with respect to the established community of natural philosophers in Europe. Almost all of the scholars who went to Europe were churchmen having the blessings of the Patriarchate. They were among the best who had mastered the amalgamation of the ancient thought together with the teachings of the Church. In their travels to Europe, however, they found Europe quite different from what the narratives and experiences of the scholars of the preceding generation had led them to expect. During the early part of the eighteenth century they found a Europe dominated by the ideas of the Scientific Revolution, with flourishing scientific communities involved in the production of original scientific work. The institutions where the Greek scholars could indulge in the all-embracing studies of philosophy, continuing the kind of education they had already acquired, were progressively decreasing. The scholars were faced with a paralyzing dilemma: if they were to become part of the community of the natural philosophers in the places where they were studying, the Greek scholars had to abandon their own tradition of religious humanism. *Being ideologically unwilling and intellectually unable to proceed to such a break, they immersed themselves in the study of the new sciences with a view to returning home and assimilating them into the curriculum of religious humanism.* A characteristic example of this attitude was the increasing desire to teach the new sciences in a manner that harmonized with the conceptions of the ancients. No wonder that almost all the books on the new theories written by Greek scholars in the eighteenth century reflected, and very often explicitly expressed, their "debt" to their ancient predecessors. This conception of an uninterrupted continuity and the perfection of ancient knowledge — a conception that was essentially adopted and promoted by the Church — was one of the basic characteristics of the Greek scientific culture during the Enlightenment.

The writings of the Greek scholars reflected three traditions, at times in conflict with each other, at times complementing each other. These were the scholastic-Aristotelian tradition, the neo-Aristotelian tradition and the tradition of the European Enlightenment. The introduction and teaching of the sciences necessarily reflected a synthesis of traditions which, quite often, obeyed ideological and political aims rather than complying with the dominant problematic of the natural philosophers of Western Europe. Of course, the Greek scholars were fully conscious of the deep influence exerted by natural philosophy on political philosophy and this was not a secondary factor in their intellectual itineraries. Finally, such an interpretative framework helps us to understand why almost every one of the scholars who had played a significant role in the introduction of the new scientific ideas in the Greek speaking regions, had written a book in philosophy or logic *before* publishing a scientific book. Physical, astronomical and cosmological writings give us an additional probe into the character of this idiosyncratic discourse that Greek scholars developed in their attempts to introduce the new scientific ideas.

During that period the main pursuit of the Greek scholars remained the articulation of a philosophical discourse, within which the new ideas of the scientific revolution would be fused with the aristotelian tradition and the Orthodox Christian ontology. In 1716, almost a century after the appointment of Theophylos Korydalleas as the director of the Patriarchal Academy, Chrysanthos Notaras (1663-1731) published his work *An introduction to the Spheres and Geography.* At the time, Chrysanthos Notaras was Patriarch of Jerusalem, succeeding to his uncle Dositheos. Both had been particularly active in consolidating the economic and political presence of the Orthodox Greeks in the Holy Lands and in opposing the claims of the Catholic Church. Chrysanthos Notaras had started his studies at the Patriarchal Academy and, later on, in 1696, he was sent to Vienna and later to Venice and to Padua. In 1700 he visited Paris where he made astronomical observations at the Paris Observatory for several months. Though he never practised as a teacher, in 1707 he proposed a program of reform for the Academy of Bucharest, which later was also used for the reforms instituted at the Academy of Iasi (1714).

Ancient Greek literature, philosophy, and the natural sciences were the main courses of that program. However, Chrysanthos Notaras proposed the teaching of philosophy according to Korydalleas' neo-aristotelian model, though he himself had already studied at European universities almost a century later. Chrysanthos' astronomy followed in Korydalleas' footprints and his aim was to advocate the geocentric system as opposed to the current theories of heliocentrism. Though he exhibited some tolerance towards Copernicus and discussed his ideas, Chrysanthos Notaras, as a staunch follower of religious humanism, pointed out that Copernicus' ideas were nothing more than a reproduction of the cosmological model proposed by Aristarchos of Samos. In the first part of his book he briefly described the main schema of aristotelian cosmology, while in the third he developed in detail the arguments in favour of geocentrism and against the rival heliocentric system. His conclusive argument had to do with the absence of any internal or external cause that could *account for* the «circular or straightforward movement of the Earth». The absence of any possible cause was sufficient proof that the Earth remained still in the centre of the universe. Chrysanthos Notaras' conclusion was philosophico-theological rather than scientific in nature: *since the Ptolemaic system could describe the celestial phenomena at least as adequately as the Copernican system could, the adoption of the former was necessitated by its agreement with the Scriptures and the senses.*

This situation changed radically after about the middle of the 18th century, when a great number of the Greek scholars became supporters of the heliocentric system. That heliocentrism found quite a few adherents was not independent of the fact that the polemics against heliocentrism were no longer particularly intense. Those who were against the heliocentric system presented the alternative cosmological systems to their pupils, and came out in favour of the geocentric system, based either on the Scriptures and/or Aristotle or, as Evgenios Voulgaris did, on recent observations which could not find evidence of stellar parallax. We should also note that the Copernican system had a peculiar affinity to the Greek thought: many authors presented the heliocentric system as originating in Pythagorean ideas. Hence, heliocentrism could be considered as part of a national spiritual heritage —a reminder that the Church continued to be the guarantor of the traditions of the Greek nation. For that reason it was not strange to see both systems often accepted as valuable hypotheses: though geocentrism was to be preferred, heliocentrism —to the extent that it had its origins in the ancient Greek thought— did not *necessarily* undermine the ontological contentions of religious humanism.

One of the adherents of heliocentrism was Iosipos Misiodax (1725 - 1800) who appears to have been fully conscious of the fact that the traditional model of the cosmos by Aristotle had not been simply subjected to some minor changes, but had been displaced by a new and dramatically different system. In his works he presented the claims of Copernicus and he also discussed the invention and role of the telescope, the discovery of the solar spots as well as Newton's gravitational theory. Iosipos Misiodax was born in Cernavoda, Bulgaria. He studied at the School of Mount Athos between 1754 and 1755 when Evgenios Voulgaris directed the school. Towards the end of the same decade he visited Venice, Padua, Hungary, and Vienna. In 1761 he translated and published Muratori's *Moral Philosophy,* in Venice. In 1765, he started his career as a teacher in the regions around the Danube where the Ottoman Court had appointed Greek governors. A committed educational reformer, his ideas were considered to be in opposition to the theology of his days; to defend himself from the attacks of his opponents, he published in 1780 his *Apologia,* in Venice. Though his most forceful contributions were in the field of political and social philosophy, his works contained extensive references to the recent attainments of the sciences.

The following year, Misiodax published the *Theory of Geography,* where among other topics he presented the various theories concerning the motion of the Earth. Though it was obvious that he was a strong defender of heliocentrism, he tried to safeguard himself: while he was in general agreement with the beliefs of most of the natural philosophers in the West, he continuously scrutinized the logical structure of their arguments. In the same text, and in trying to moderate reaction against the presentation of heliocentrism, he wrote that heliocentrism was put forth so as «to invent new means to imitate the Pythagoreans», thus attempting to dress the new theories with the respectable cloth of antiquity.

Misiodax recognized that the place of man within the cosmos had been altered, and as a result, the order of values had changed as well. At the same time, he was aware of the course which the newer sciences had charted in modifying the traditional world-view. Though Copernicus was the «glorified rejuvenator of real Astronomy», the breakdown of the aristotelian universe was caused by the use of the telescope, the discovery of sunspots, and the study of comets, which showed that the universe was uniform and homogeneous, and that there was no physical distinction between the world above and below the Moon. The culmination of his ideas was considered to be the combination of new mathematical thought with the experimental tradition within the framework of the Newtonian synthesis.

Misiodax ascribed to mathematics an educational merit which he had not ascribed to logic and metaphysics. In fact, he had accused Aristotle for undermining the interest in the study of mathematics and expressed his admiration of Galileo, Descartes, and Newton because, in his opinion, they restored the ancient respect for mathematics; he thus considered the teaching of mathematics together with experimental physics to be of great value for a modern educational curriculum.

In his work *Theory of Geography* he presented the arguments for the heliocentric system together with the claims of the adherents of geocentrism. Although he claimed to keep equal distance from both views, his way of organizing and presenting the arguments clearly indicated his preferences. He started his analysis with a programmatic distinction between science and the Holy Scripture, and he was led to the rejection of all theological arguments concerning the structure of the universe. He, then, reconsidered the issue of the absolute reliability of the senses, which was a strong cognitive bulwark of the aristotelian thought; at the same time, he rejected the premise that phenomena which take place below the Moon are distinguished from the ones which occur above it, and he expressed his agreement with the idea that nature is homogeneous. His arguments for the Earth's motion were completed with an appeal to Kepler and Newton, whose laws, as he stressed, confirmed and validated the heliocentric hypothesis. Misiodax adopted the ideas of the new sciences without having to devise detailed arguments against aristotelianism. He considered the break with the aristotelian cosmology to be the end of a whole era and established his proposals for educational reform upon the undisputed acceptance of the new image of the cosmos as put forth by contemporary science.

Another eminent scholar of the time was Evgenios Voulgaris (1716-1806), one of the more intriguing personalities of the Greek Enlightenment. He was born in Corfu and died at the age of ninety at the court of Catherine the Great. He studied in Corfu under Vikentios Damodos, an important scholar of the period. He continued his studies in the School of Ioannina (a wealthy commercial town of western Greece) under Athanassios Psalidas. After he became a priest, in 1737 or 1738, he went to Italy in order to study theology, philosophy, european languages and natural sciences.

In 1742, he became director of an important school in Ioannina. There he was involved in a public dispute with Balanos Vassilopoulos, who was the director of another high level school of the district, regarding their respective curricula – Voulgaris arguing for the introduction of natural philosophy. From 1753 to 1759 Voulgaris was director of the School of Mount Athos, where he worked to upgrade the level of studies. There he taught philosophy as well as mathematics. Though he was considered to be one of the most eminent teachers, his strong adherence to the new ideas provoked reaction from the religious hierarchy of Athos, and he was forced to abandon the school in early 1759. He then moved to the Patriarchal Academy, and in 1761 he permanently abandoned his educational career. The presence of Voulgaris in the various schools of the Greek speaking regions gave rise to antagonisms because of his adherence to the new scientific ideas and his self-asserting personality.

Voulgaris wrote and translated many books on a wide variety of subjects, mainly for use in his own teaching. Most of these books remained unpublished or were published many years after the end of his educational career, while he was living in Russia. Like most scholars of his time he launched his intellectual activity with a book on *Logic* (Leipzig, 1766). He continued with the publication of *Elements of Metaphysics* (Venice, 1805), *What Philosophers Prefer* (Vienna, 1805) —an amalgam of the recent ideas of the sciences and Philosophy— and *About the Universe* (Vienna, 1805) where he discussed the contemporary astronomical theories. It is interesting to note that he also translated —though not published— works of Locke, Wolff, Du Hamel, and Pourchot.

Although Voulgaris, an ordained priest, was one of the most systematic initiators of the new scientific ideas into the Greek speaking world, he remained throughout his life an adherent of the idea that the gnoseological authority of the Scriptures was much more valid than any other cognitive approach to the world. In this sense, Voulgaris typified the case of a Greek scholar who assimilated the attainments of the Enlightenment by incorporating them in a discourse within the restricted framework of religious humanism. Voulgaris seems not to have recognized the radical reorientations in the study of nature brought about by the introduction and legitimation of experimental procedures. He was trying to combine the new theories with the teaching of the ancients, which he strongly believed to be the foundation for any modern knowledge. He refused to acknowledge the crucial role of experiment and mathematics in the new scientific *discourse* developed among the natural philosophers in Europe. In Voulgaris' view physics should derive its conclusions mainly through reasoning. It is true that in his translation of Segner's *Elements of mathematics* he underlined the necessity of algebraic knowledge for the study of the world and for the proper understanding of physics. His perception of the proposed connection of mathematics with physical inquiry, however, was restricted to a mentalist context, and in his subsequent work he tended to connect the use of mathematics with philosophy, following the exemplar of the ancient Greeks. As will be discussed later, the case of Veniamin Lesvios —another eminent scholar of the Greek Enlightenment— was similar to that of Voulgaris: Lesvios suggested that the application of geometry in astronomy gave the latter the quality of a science; but, although he acknowledged the contribution of analytic geometry in the formation of the modern scientific discourse, he considered it to be difficult as well as lacking in elegance and he advocated Euclidean geometry as appropriate for the education of young people.

Voulgaris' work *About the Universe* was written for teaching purposes and contained his main astronomical views. Though it was published in 1805, it is presumed to have been completed before 1761, the year he abandoned his educational activities. When the book was published, the 89 year-old Voulgaris was a recluse at a monastery hardly communicating with anyone and, thus, it was not at all clear whether he was even asked to acquiesce in the publication of the book. It was mainly an amalgam of the work of various European philosophers. Voulgaris did not hesitate to recognize that the Ptolemaic system was obviously contrary both to observations and common sense. That, however, did not mean that he espoused the truth of heliocentrism. He mentioned a number of arguments expressing the lack of any sense-experience for the revolution of the Earth. His main argument, however, was of gnoseological nature and had to do with the validity of the Scriptures: though their main target is the salvation of the human soul, they also accommodate some natural truths which are able to reinforce moral teachings and reveal Divine Providence; and though the movement or stillness of the Earth is irrelevant for the salvation of the soul, we are obliged to accept the divine assurance of the Earth's stillness as the most reliable.

The Ptolemaic system was inadequate and heliocentrism involved cognitive obstacles that prevented Voulgaris from accepting it; he thus preferred a third interpretation as the most reliable, namely the system proposed by Tycho Brache. Voulgaris was the only scholar in the Greek speaking world who embraced the Tychonic system in the middle of the eighteenth century. His argumentation, however, was not simplistic; he was fully acquainted with the issues involved. His main objective was to confirm that the Tychonic system was in accordance with astronomical observations, that it interpreted a host of phenomena that Ptolemy's system could not adequately explain, and that it was technically at least as valid as the heliocentric system. Then, since it kept the unmoving Earth at the center of the universe, the Tychonic system was consistent with the Scriptures and, on that ground, preferable.

In 1794, the Fanariot Panagiotis Kodrikas translated Fontenelle's *Entretiens sur la pluralité des Mondes*. This semi-popular book discussed the idea of an infinite universe and the plurality of worlds. The translation included many notes by Kodrikas. One would not have expected that a translation of a book written nearly a hundred years earlier would be a high priority. Yet, Kodrikas chose to translate it so that he could include his extensive "explanatory" notes which in fact amounted to a parallel text. The translation provoked a strong attack against the followers of heliocentrism, and finally was condemned by the Church — not because of the ideas of Fontenelle, but because of the translator's comments which were in explicit conflict with the traditional values of religious humanism: he did not hesitate to write against "those who due to superstitious ignorance do not acknowledge the established truths of the Copernican system." With Kodrikas' translation we have, for the first time, definite recognition of the role of the founders of modern physics (with the exception of Galileo who does not seem to be so dear to the Greek scholars of the time): it is noted that while Copernicus had, in fact, put the Sun in the centre, he had not changed the structure of the cosmos, and that Descartes had broken away from the Aristotelian world-view; finally, the significance of Newton's synthesis is brought to surface.

The French Revolution did not sit well with the Fanariots' political agenda. Many of them considered the Revolution and its consequences as endangering their prospects of increasing influence within the Ottoman Empire. As the French Revolution was more and more projected as the realization of the political and social ideas of the Enlightenment, the Fanariots' belief in and attachment to the ideas of the Enlightenment started to weaken. Also, as the anticlericalist positions of the Revolution were associated with the spirit of the Enlightenment, many scholars — who, as we stressed, were men of the Church — became less and less willing to be identified with the ideas of the Enlightenment. Naturally, we are not talking of a radical change which was adopted by all concerned: quite a few scholars, especially teachers, continued to remain strong adherents of the new scientific ideas. However, we do stress a change of heart among many scholars in their strong backing of the ideas of the Enlightenment, which, as a result, allowed a greater leverage to those in the Church who were strong opponents of these ideas from the very beginning.

This rather mixed situation and change in the mood of the scholars was quite evident in attempted changes in school curricula. That is displayed in a very typical way at the *Megali tou Genous Scholi* (a continuation of the Patriarchal Academy). During 1778-1801, the director of the school was Sergios Makreos, who remained faithful to the traditional educational system of the school. He reacted against the proposed reforms and was not even willing to include in the curriculum the books of Voulgaris, who was Makreos' teacher and for whom he felt great respect.

Makreos was born around 1740 and died in 1819. His education was entirely within the Greek speaking areas of the Ottoman Empire. He attended classes at the School of Agrafa (a mountainous region of central Greece) and then he went to the School of Athos under Voulgaris. He completed his studies under Voulgaris at the Patriarchal Academy in Constantinople. During his directorship he taught Aristotle according to Korydalleas' system, and though he was proud of being Voulgaris' student he refused to bring into the curriculum the new sciences and philosophy, because he believed that they would lead to the breakdown of traditional social values.

In 1797 he published *A Trophy from the Greek panoply against the followers of Copernicus*. Makreos seems not to have been acquainted with the course of events and conceptual changes which led to the final Newtonian form of the heliocentric theory. By attacking Copernicanism he attacked all the new astronomical theories —among them the infiniteness of the universe and the inhabitation of the stars. Makreos' arguments against heliocentrism were not predominantly "scientific"; instead he stressed the social repercussions of each model: the overthrow of the traditional hierarchical and static world expressed by geocentrism would entail the downfall of social hierarchy.

Makreos attempted through a number of syllogisms to discredit Newton's theory of gravitation. He considered that the motion of every body was determined by the outcome of two forces, the centripetal and the centrifugal. He discussed two cases. The first was when both forces are applied to the same body specifically to the Earth. According to the Aristotelian viewpoint centripetal force is natural whereas centrifugal motion is violent. But again according to Aristotle it is impossible for a body to execute natural and violent motion at the same time. Hence there is a contradiction. In the second case one of the forces is applied to the Earth and the other to the Sun. The net result of this situation would be for the Earth to be attracted and repelled by the Sun at the same time. According to Makreos this has the following consequences. Either the Earth will move in a straight line and there will be no reason for it to revolve or, if the two forces are equal in strength, then the Earth will be motionless for ever. Makreos concludes that the law of universal attraction cannot justify the heliocentric system. It is important to underline the qualitative character of his arguments as well as his disregard of the law of inertia. Of course, Makreos had a rather shrewd strategy: he used the most «quantitative» of Newton's laws and showed that on Aristotelian premises it is self-contradictory and cannot lead to what Makreos considered as unquestionabe qualitative characteristics of the cosmos.

The way he developed his arguments clearly shows his intention to work within a pre-determined cosmological schema. At the same time, he presented an analysis of the gnoseological preconditions for heliocentrism, which, in effect, comprised the main part of his book. He adopted the main points of the Aristotelian worldview and rejected the image of an infinite, homogeneous world implied by the new astronomy. He tried to restore the validity of sensory experience and proposed simplicity as the criterion for the correctness of astronomical theories. Furthermore, he questioned the validity of astronomical observation together with the reliability of "mathematical instruments". He claimed that we cannot determine what is happening in the sky from what we observe from beneath the Moon, because these are two different realms with two different classes of phenomena.

Makreos was committed to defending the aristotelian world-view within the context determined by the Orthodox Christian faith. A difficulty here was that Aristotle's views on the eternity and self-motion of matter could be interpreted as rendering the act of Creation unnecessary. For this reason, part of Makreos' task was attempting to explain that matter is created and it is passive. Creating it and putting it into movement demanded the mediation of the highest power and of the eternal energy of God. By placing cosmology within this context Makreos was stressing, at the same time, the cognitive limits of man. There is no way for knowledge to surpass the limit imposed by the relationship between humans and their Creator. Therefore, knowledge should always be subjected to the truth of the sacred Scriptures. Almost two hundred years after Galileo and his telescope, Makraios remained an aristotelian by totally rejecting the possibility of even a partial updating of traditional cosmology.

* * *

From the middle of the 18th century, economic well being of the Greek communities within the Ottoman Empire with the accompanying social transformations brought about a number of changes in the educational system. The reception and appropriation of the new scientific ideas went on within an environment of social unrest and ideological confrontations. One cannot talk about educational reform, since the attempts were local initiatives rather than a centrally dictated policy to be applied to a homogeneous educational system. While in the seventeenth and at the beginning of the eighteenth centuries all schools were religiously oriented, the coming years saw the emergence of schools whose curriculum could cater for the social and political agendas of the merchants or the Greeks involved in the administration of the Ottoman state. The systematic introduction of the sciences was reinforced by renewed faith in man's ability to acquire knowledge of the world with his own means, and all these found support in the expectations of the assertive merchants and in the political ambitions of the Greek officers of the Danube region.

Within this context, there was a gradual re-definition of the teachers' role. The image of the teacher-priest whose work was a religious mission gave way to another kind of scholar: the great majority of these teachers were priests, but their educational agenda became more secular and their actual work tended to be more «professional». The scholastic and grandiloquent teaching of the works of the Fathers of the Orthodox Church, of ancient Greek literature and of Aristotle, gave way to a curriculum determined through negotiations with the community which had established and catered for the schools. Teaching began to serve the social, political, and ideological «priorities» of these communities. These changes strengthened the relative autonomy of the scholars from the Patriarchate and reinforced their role as independent scholars.

At the end of the eighteenth century, the introduction of Enlightenment ideas in the schools became the subject of social negotiation. For many Greek scholars the European nation states represented a model, while at the same time they provided the ideological background for a new political discourse in the Greek speaking world. The methods for the introduction of knowledge were changing, because, as the scholars were progressively realizing, the knowledge to be introduced was of a new kind. Teaching had to become pleasant and attractive in order to cultivate curiosity and independent thinking. Visual means supplemented texts: maps, globes, experimental instruments, experimental demonstrations themselves became the pride of teachers. Constantinos Koumas (1777-1831) claimed to be the first teacher who conducted physical and chemical experiments at the Philological Gymnasium of Smyrna, in 1812. Philosophy became part of the educational process as a source of social ethics and good reasoning. Thus, observation, experience, the cultivation of good reasoning, and the fight against superstition became the main educational objectives, necessary for combatting ignorance and bringing progress according to the promises of the new sciences.

The cultivation of the sciences was aiming at the integration of the Greek speaking populations into the group of European nations, its other objective being the strengthening of the secular power to counterbalance the Church. At the same time, the natural sciences appeared to be the answer to the social demand for craftsmen-technicians, and for improving the efficiency of the merchants' dealings with their counterparts in Europe. The publication of commercial handbooks and the establishment of progressive schools were also expressing that trend, with the clear objective, at least of some merchant groups, of undermining and discrediting the traditional educational system.

During the whole of the eighteenth century and until the Greek Revolution of 1821, there was no branch of the natural sciences —with the exception of medicine— organised into a distinctive cognitive field with institutional autonomy. The scholars who were dealing with the newest scientific ideas were differentiated on the basis of their general ideological and political affinities; only medical doctors tended to become a separate profession. The work of the rest of the Greek scholars had social aims within the framework of the interests of various social groups and this is the reason why scientific discourse appeared in the context of a more general political agenda. The ideas that had originated during the scientific revolution were viewed more as an educational activity responsive to certain social demands or contemplative quests, and less as a method for practical research. They were knowledge to be acquired, not a method for producing new knowledge. In the Greek speaking regions, we cannot trace a trend whereby scientific practice is transformed into a socially structured activity having as its main element the empirical—let alone experimental—study of nature.

Nevertheless, towards the end of the eighteenth century the number of published scientific books began to increase. Greek scholars started writing and translating a large number of scientific works to be published in cities like Venice and Vienna. A common practice was the dedication of the works to eminent persons of the Greek speaking world. Dedication was part of the politics of patronage. On the part of the writer this politics aimed at legitimating his work and his ideas expressed through it. The most important element of this practice, however, was the promotion of specific social values: the cultivation of literature and the arts, of virtue and piety, as well as the dissemination of political visions for the benefit of the nation.

A point to be stressed is that the scholars seemed equally interested in legitimating the content of their own work and in a more general social programme within which their work would find its place.

The readership for these scientific works became progressively more diversified. In the middle of the eighteenth century, scholars like Nikiforos Theotokis and Evgenios Voulgaris were writing for specific cultivated audiences. Their books published in the decade of 1760 addressed their readers as «friends of the sciences». Towards the end of the eighteenth century, however, the authors started addressing their readers in a more general way, without attributing specific qualities to them. The phrases «to the Greeks» or «Philhellenes» also appeared at this time. The word «Greek» together with a reference to «nation» still retained the meaning of «learner of (ancient) Greek» or «educated person» as it had in the middle of the century; a widening of the meaning, however, was now taking place; the word acquired cultural connotations related to the collective consciousness of the Greek speaking Orthodox Christians of the Balkans. This widening brought about changes in the idea of science, its role, and its cultivation. «To the reader» was a common address in the prefaces of scientific books of the time suggesting that expectation of the «uprising of the nation» could also be helped through scientific education. Such an address implied the idea of education as a key element in the concept of *citizen*. In this situation, the «development» of science meant its spread to as many people as possible: Though not everybody was capable of practising science —since that was a matter of specialization— everyone did have the potential to acquire scientific knowledge for the enlightenment and happiness of the nation. At the beginning of the nineteenth century, when the question was raised as to who would have the authority to decide about the soundness of the different scientific conclusions, Greek scholars gave an answer characteristic of the way they had perceived the ideas of the scientific revolution: While in the West the newer scientific discourse was already formulated as a network of regulatory principles handled by a structured scientific community, the Greek scholars considered the general literate public to be a legitimate judge of scientific knowledge. The «*principles* of science» were considered to be sufficient knowledge for anyone to take part on an equal basis in a discussion with the natural philosophers in the west, since exploration of natural issues demanded nothing more than good reasoning and common sense.

The ambivalent attitude towards the Enlightenment after the French Revolution was registered in the various schools of the Greek speaking areas of the Ottoman Empire. One example was the school at Kydonies in Asia Minor. The man who played an important role there was Veniamin Lesvios (1762-1824), who studied Mathematics and Physics at Pisa and Paris during 1789-99. In Paris, Lesvios met Adamantios Korais (1748 - 1833) —the «theoretician» *par excellence* of the Greek Enlightenment— and was deeply influenced by his views. Lesvios proceeded to a number of reforms and during his directorship (1802/3-1812), the school acquired the reputation of the best school for science. There, Lesvios taught the new Physics and —something quite unique— the heliocentric system *per se*; he also taught Philosophy and Mathematics. However, during his years at Kydonies, Lesvios was accused of introducing dangerous innovations through his scientific teaching, and rejecting divine incarnation. He was, thus, forced to defend his own religious orthodoxy to the Patriarchate, though he was not asked to deny his scientific beliefs. Living, however, in a period in which the Ecumenical Patriarchate had officially expressed its opposition to the new ideas of natural philosophy —because of their ideological and political implications— he did not succeed in defending the orthodoxy of his beliefs, and he was condemned by the Holy Synod. In 1819, he left Kydonies and went to the Peloponese, to take part in the uprisings of the Greek Revolution.

Before proceeding with the examination of Lesvios' physical philosophy, we should note the emergence of a distinct anti-European trend in the early years of the 19th century. Athanassios Parios was the most characteristic representative of this trend. Parios had spend some years at Mount Athos and had then become a teacher of Greek. He taught at the school of Chios, an island near the coast of Asia Minor, in the same years that Veniamin Lesvios was teaching at the school of Kydonies. Because of his extreme conservatism, many scholars of that period attacked him and students gradually started to abandon the school, moving to the school of Kydonies. Parios had some general knowledge of physics which he had acquired from the classical treatises and the commentaries on Aristotle's physics. He was the writer of a polemical book called *Response* published in Trieste in 1802. The full title of this work speaks for itself «Response to the frenetic zeal of the philosophers who come from Europe; exposing the vanity and folly of their lamentable efforts exerted upon our Race and teaching which is the real and true philosophy. To which is added a salutary admonition to those who recklessly send their sons to Europe on business». The Christian, declared Parios, should not examine the secrets of the creation of the material world, because in this way we are able to reach only hypotheses and not proven conclusions. For that reason the mathematical sciences are the source of godlessness because they create an illusion of certitude.

Let us now discuss the work of Lesvios and especially his *Πανταχηκίνητον* (The All-Mover). This «theory» is a paradigmatic case in support of our main contention. It is a work whose structure and argumentation is incommensurate with the dominant scientific problematic of the period and, at the same time, a characteristic example of an attempt to form an alternative scientific discourse. It was never published but was systematically taught; Lesvios' manuscript, in which he developed the theory and discussed physics and astronomy, dates from a few years before 1800.

Lesvios had serious objections to the validity of Newton's first law. He could not accept that bodies, left to themselves, would continue to preserve their kinetic identity. He maintained that the motion of bodies left to themselves would run down. In other words, he disagreed with what had been accepted as the constitutive aspect of the new physics: The principle that force is necessary *to change the direction* of motion. For Lesvios, it was the initiation of motion that required force. He went on to explain the revolution of the planets through the assertion of an effluvium (Πανταχηκίνητον) which is emitted from and absorbed into the bodies in proportion to their mass. From this schema it followed that the body with the largest mass (the sun) must be at the centre. The equilibrium of forces as a result of absorbed and emitted effluvia maintains the stability of the distance of the planets from the sun, the rotation of the sun on its axis guarantees the revolution of the planets around the sun, and the difference in intensity between the effluvium of the sun and that of a planet, when these «meet» and create a kind of vortex near the planet, gives the latter a rotation on its axis.

Newton's first law was not merely a synthesis of the various issues related to inertia. Equally important, it formed part the context of consensus about the ways the new physics would be practised. The first law dictates the study of the changes in the direction of motion and precludes the search for the causes of motion. Even if at the beginning of the 18th century natural philosophers did not all agree on the range and the character of the legitimate questions to be asked within the framework of the new physics, by the time Lesvios was formulating these ideas there was no doubt about the kind of questions natural philosophers were allowed to ask. Lesvios' problematic and his methodology made up a program for understanding the metaphysical foundations of physics; in this respect, Lesvios' agenda was much closer to the programs of Leibniz and Kant. Certainly his difficulties in accepting action-at-a-distance were quite decisive in formulating his theory of Πανταχηκίνητον. But he must have been also influenced by the generally favourable climate in Europe concerning the heuristic value of the imponderable fluids. But even if we grant that Lesvios' belief in these fluids may not have been undermined by the developments in chemistry and the recent explanation of oxidation, the way he developed his theoretical schema was qualitative, and aimed at explaining what was already known and observed; nowhere did he indicate the possibility of either a new phenomenon to be associated with his schema or a quantitative prediction which could be measured.

Lesvios had been educated in Europe and in his writings we witness a quite systematic knowledge of what the state of science was. How are we to understand what he was attempting to do? How can we understand the rejection of Newton's laws and his theory of gravitation and, at the same time, the adoption of the heliocentric system? What should we make of his preference of demonstrations rather than of experiments? And how are these to be contextualized when we know that he was one of the ablest teachers, by no means a charlatan with this idiosyncratic universal theory?

Lesvios attempted to articulate a form of discourse with the following characteristics. Based on metatheoretical elements of the dominant schemata in physics (the imponderable fluids and/or heliocentrism) and on a still unanswered difficulty about Newton's theory of gravitation (action at a distance) he proceeded to the formulation of a philosophical system where the foundational principles would lead to the explanation of as many phenomena as possible. Thus it is *not* strange that he rejects Newton and adopts heliocentrism. It is, also, *not* strange that he, and especially his student Kairis, extend his theory to include human feelings as well —in fact, such a theory must be able to be extended in such a direction. Lesvios developed his theory of Πανταχηκίνητον within a framework of what *he considered* as physics and astronomy and not as part of his metaphysics. In other words, Lesvios' physics is neither a popular or didactic presentation nor a piece whose purpose is to inform about the developments in the west. Viewed as an alternative theory *within* the framework of the scientific discourse as formed in the west, Lesvios' agenda has no legitimation whatsoever. But if it is seen as an attempt to propose an *alternative to* the (western) scientific discourse — a philosophical rather than a scientific discourse — then his whole program appears considerably less idiosyncratic.

Lesvios' work is a typical case of *appropriation* of the new scientific ideas into the cultural milieu of the Greek speaking regions. His work cannot be interpreted if the dominant methodological tool and historical category we use is that of *transmission*. It is, no doubt, the case that many ideas had been introduced into the Greek speaking regions by ways which can be perfectly well understood through the use of the concept of transmission. These are the easy and straightforward cases. However, the effectiveness of a methodological tool is measured by the possibilities it provides for the understanding of what appears to be exceptional — in this case, what appears to be a capricious, superficial and «less than scientific» theory by a well educated scholar. Thus, during this period a number of issues were reformulated in order to be appropriated within a context determined by a number of philosophical needs, ideological outlooks, and political imperatives. The appropriation of the new scientific ideas called for the formation of the necessary discourse which appeared to reflect the *network of constraining localities*.

During the time that Veniamin Lesvios was teaching at Kydonies, another eminent scholar was the director of one of the most famous schools in the Greek speaking regions. This was Constantinos Koumas, director of the Philological Gymnasium of Smyrna in the western coast of Asia Minor. As director, he was responsible for the teaching of scientific courses. The Philological Gymnasium of Smyrna was one of the most important educational centres of the Greek speaking regions in the last period of the Ottoman Empire; thus, it is interesting to see how the role of sciences in the social context of the time was reflected in the curriculum of the school.

Constantinos Koumas was born in 1777 and was one of the Greek scholars who practised teaching as a profession and not as a part of their religious vocation. He was very interested in mathematics from his early years. After a short period of teaching, he went to Vienna, in 1804, where he completed his studies and published various scientific works. Koumas had a doctorate in Philosophy and Fine Arts at the University of Leipzig and was a member of the Royal Academies of Berlin and Munich. In 1809 he was invited to Smyrna to take over the Philological Gymnasium; he remained there up to 1814, and then went to Constantinople to take over the directorship of the Patriarchal Academy.

Under his directorship, the school in Smyrna became famous, especially because of his methods of teaching the physical sciences: the «mysterious instruments» used in classes attracted a great number of wealthy Greeks, who sent their children to study at the Philological Gymnasium of Smyrna, while the Evangelic School, another important school in Smyrna with theological orientation, gradually declined. Nevertheless, in the Philological Gymnasium itself, a current of opposition to the introduction of scientific ideas and the secular orientation of the curriculum gathered momentum, especially after Koumas' departure. The scholars who advocated the conservative policy of the Patriarchate reacted to the liberal policy of the merchants, who continued to support the introduction of the innovative ideas in the intellectual life of the Greek speaking populations. As a result, in 1819, during a major political disturbance in Smyrna, a violent crowd set fire to the Philological Gymnasium.

The teaching of the sciences remained the main axis of the curriculum at the Philological Gymnasium of Smyrna. A key argument for the introduction of scientific courses was the need for scientific knowledge to return to its birthplace. Koumas in his translation of Pierre August Adet's *Chemistry*, in 1808, defined chemistry so broadly, and made such an *ad hoc* interpretation of Aristotle's texts, that he was able to conclude that the ancient Greeks were the real initiators of current chemistry. For Koumas, in fact, it was impossible for ancient Greeks, who had developed every other science and art, not to have developed chemistry as well.

In 1812, Koumas published *A Synopsis of Physics* in which he also developed his educational program: «According to the ancients, Philosophy was divided into three parts, Logic, Ethics, Physics. Every kind of science and art is reduced to these three genres of Philosophy. Whoever intends to lead a good life in society should not ignore any of these parts». In defining Physics, however, he was more interested in the development of a visual teaching method than in teaching a method for scientific research. Physics was the science «that teaches us about the phenomena, as well as the reasons or forces which cause them». Though experiment and observation were considered as the key methods of research, the whole concept was closer to an Aristotelian qualitative interpretation of the phenomena than to the quantitative study of nature and the derivation of mathematically formulated conclusions. It should be pointed out that the school in Smyrna acquired its fame — among other things— because of the physical and chemical demonstrations that were conducted by Koumas, who also supplied the school with maps and globes. But these were not experiments related to any kind of original scientific research; they were rather repetitions of the demonstrations of phenomena, which had already been studied by the natural philosophers in the west. Thus, though Koumas stressed the usefulness of experimental research, in his pedagogical practice the separation of experimentation from mathematics, and from the quantitative evaluation of its results, dissociated it from its specific heuristic role within the scientific discourse developed in Europe.

At the time that Koumas published *A Synopsis of Physics*, Constantinos Vardalachos, another important scholar, published his *Experimental Physics*. This work consisted of a collection of his analytic notes for the courses he had been teaching in the sciences at the Academy of Bucharest. Vardalachos makes a distinction between mathematical physics, «which is proven by geometry and calculation» and experimental physics «which is proven by the phenomena [i.e. observations]» —justifying, in a way, the elimination of mathematics from his teachings. In the beginning of the nineteenth century, even though the introduction of physics in the curricula of the modern Greek schools was considered one of the main intellectual innovations, the almost exclusive use of qualitative interpretations allowed it to be kept within an aristotelian context. The notion of experiment and observation was unrelated to the way experiments were being conducted by the scientific community in the West during the same period. For Greek scholars, experiments were demonstrations intended to motivate students and to convince them of the validity of the qualitative interpretations concerning the origins of various phenomena. The use of experiment for the discovery of new phenomena and/or for the quantitative survey of the natural world, as was the case with the western natural philosophy, was outside their scope.

FINAL COMMENTS

Let us summarize some of the salient aspects of these developments.

It appears that a standard approach, with the emphasis on understanding the formation and function of social institutions such as patronage and the academies, is inadequate in the case of the Greek speaking populations during the 17th and, especially, the 18th century. We want to understand the ways the new scientific ideas were introduced and established in a region which was part of the Ottoman Empire. The jurisdiction of the Church over educational matters, its initiatives for sending scholars to Europe to be educated and the kind of dynamics created as the intended and, most interestingly, the unintended result of their scholarly work — whether by writing books or teaching — all need to be assessed within the overall particularities of the Greek case. A number of complicated issues will also have to be taken into consideration. The ambivalence of the Church towards the shifting philosophical allegiances and the ideological orientations of the scholars; the relations of the Church with the Ottoman administration; the relations between the ecumenical Patriarchate of Constantinople and the other (autonomous) Patriarchates each facing different problems of their own (e.g. the Moscow Patriarchate and the initiatives for modernization by the new ruling classes of the 18th century); the relations of the Orthodox Church with the Holy See and the Protestant world; the interests of the prominent and rich laymen at Constantinople, often in conflict with those elsewhere in the Balkans.

It was accepted by all that the Patriarchate had absolute responsibility in formulating the long term educational policies and articulating the ideological agenda for a synthesis between hellenism and Orthodoxy. This did not mean that the ensuing developments went smoothly as implementations of the original programmatic directions. There appeared many different trends, each claiming ideological or political leadership of this process aimed at preserving religious identity and inspiring national consciousness. These trends were at times in conflict with each other and at times complementary. Scholars following the scholastic aristotelian tradition coexisted with neo-aristotelians. Scholars adopting the ideas of the Enlightenment came into conflict with those who viewed these ideas as undermining the conditions for religious and ideological survival. The introduction of the sciences and their subsequent teaching necessarily reflected a confluence of all these trends. The glorious developments of the new sciences in western Europe became an interesting but expected corroboration of the programmatic declarations of Aristotle. Social groups who found confidence in the ideas coming from Europe for their political future, turned against the ideas of the Enlightenment after the French Revolution. Issues related to national consciousness of the Hellenic population became separable from issues related to theological questions; religious humanism could no longer contain the antagonisms. The Patriarchate reflected and conditioned these changes. Progressively it became less receptive to ideas and policies that it had welcomed about two centuries earlier. But there again, it had mostly achieved what it had set out to do.

One of the difficulties in trying to analyse the newly emerging community of scholars in the Greek-speaking regions has to do with the relative lack of consensus among the scholars as to the *constitutive discourse* of the community. The study of the emergence of the scientific community in the various countries of Western Europe deals with the ways a group of people managed to reach a *consensus* as to the discourse they were to use in discussing, disputing, agreeing and communicating their results in the new field. In the Greek speaking regions, from the first decades of the eighteenth century until well into the nineteenth century, the discourse that the scholars developed was substantially different from that of their colleagues in Western Europe. The (expected) social role of the scholars and their ideological prerogatives legitimated a discourse which was predominantly philosophical. Furthermore, there appear to be additional reasons for the emergence of such a discourse. Firstly, there were neither internal nor external factors to precipitate a crisis with aristotelianism and, therefore, no need to reformulate, let alone initiate, a break with aristotelianism. Secondly, the dominant mode the scholars wished to establish was a kind of logic with had strong ethical implications related to the rules of correct argumentation. Thirdly, although these scholars appeared quite sympathetic to experiments, what they considered to be experiments was hardly different from demonstrations. The emphasis, usually indirect but often explicit, was about the use of the new material for (re)shaping philosophical arguments. Most importantly, there was a lack of emphasis concerning the crucial relation between theory and experiment. It is quite remarkable that in almost all the books where there is mention of experiments the emphasis is on observation and (qualitative) results, rather than on the process of measurement and dealing with numbers. In more than one place one finds passages to the effect that "rational thought is not less effective than experimental results".

The introduction of the new scientific ideas in the Greek speaking societies was a process almost exclusively directed to their appropriation for educational purposes. The apparent aim was to modernize the school curricula, but this did not mean a neutral attitude as to the possible ideological uses of these new ideas —especially the need to establish contact with the heritage of ancient Greece. The appropriation of the new scientific ideas necessarily involves a remade discourse which reflects the *network of local constraints*. As we have attempted to show, the appropriation of ideas refer to the ways devised to overcome cultural resistance and make the new ideas compatible with local intellectual traditions. Hence, understanding the *character of the resistance to the new scientific ideas* becomes of paramount importance. In the case of the Greek speaking regions the issue of resistance cannot be discussed independent of the character of the break with the ancient Greek thought. The ideological and political contingencies of Christian societies under Ottoman rule during the Enlightenment, together with the dominance of the Greek scholars in the Balkans, called for an emphasis not on the break with the ancient modes of thought, but rather, on *establishing* the continuity with ancient Greece. The Greek scholars saw the new developments in the sciences in Europe as evidence of the triumph of the programmatic declarations of the ancient Greek thought with its emphasis on the supremacy of mathematics and rationality, rather than a break with the ancient mode of thinking and the legitimation of a new way of dealing with nature. The developments in the sciences were not viewed as an intricate process which among other things involved a break with Aristotle, but rather, as developments which came to verify the truth of the pronouncements of the ancients. In addition, there were differences resulting from the respective overall social functions of the scholars in the centre and the periphery. But the development of such a discourse was also suitable for supporting the overall political agenda. The problem under consideration here was the introduction of the new scientific ideas to a national community which was under occupation and which had no national state institutions of its own. This is a very unusual situation: in the absence of national state institutions, the community lacked the conditions which would allow the effectiveness of the educational system and of the training of students in these sciences to be socially assessed. Lacking such a corroborative framework where the usefulness of these sciences would be under continuous vigilance, ideological and, in fact, philosophical considerations became the dominant preoccupation of the scholars. Hence, the embedding of all these new ideas within a philosophical context strongly at variance with that of the European scholars became an aim in itself; there was no other sense in which the new ideas could be legitimated.

Department of History and Philosophy of Science, Athens University, Greece

SELECTIVE BIBLIOGRAPHY

Argyropoulou Roxane D., "Traductions en Grec moderne d' ouvrages philosophiques (1760-1821)" *Revue des etudes sud-est Europeennes*, X, 1972, 363-372.

Cavarnos C., *Modern Greek Thought*, Belmont, Mass., 1969. The emphasis is on the philosophical ideas of the 17th and 18th centuries.

Batalden Stephen K., Catherine II's Greek Prelate. *Eugenios Voulgaris in Russia, 1771-1806*, East European Monographs, Boulder, NY, 1982.

Demos Raphael, "The Neo-Hellenic Enlightenment (1750-1821). A general Survey", *Journal of the History of Ideas* 19, 1958, 523-541.

Dimaras C. Th. *La Grece au temps des Lumieres*, Geneve, Librairie Droz, 1969.

Henderson G.P. "Greek Philosophy from 1600 to 1850", *The Philosophical Quarterly* 5, 1955, 157-165.

Henderson G.P., *The Revival of Greek Thought 1620-1830*, State University of New York Press, Albany, 1970.

Herig Gunnar, *Okumenisches Patriarchat und europaische Politik 1620-1638*, Franz Steiner Verlag, Wiesbaden, 1968.

Kitromilides, P.M., *Enlightenment, Nationalism, Orthodoxy: Studies in the culture and political thought of south-eastern Europe,* Variorum, Aldershot, 1994.

Pallis A.A., *The Phanariots. A Greek Aristocracy under Turkish Rule,* London 1951.

Papaderos Al., *Metakenosis. Das kulturelle Zentralproblem des neuen Griechenlands bei Korais und Oikonomos,* Mainz 1962.

Tsourkas Cl., *Les debuts de l'enseignement philosophique et la libre pensee dans les Balkans. La vie et l'oeuvre de Theophile Corydalee (1570-1646)* 2nd ed. Thessaloniki, 1967.

Voumvlinopoulos G.E., *Bibliographie Critique de la Philosophie Grecque depuis la chute de Constantinople a nos jours 1453-1953,* Athens, 1966.

BIBLIOGRAPHY IN GREEK

Αγγέλου Α., *Πλάτωνος Τύχαι (Η λόγια παράδοση στην Τουρκοκρατία)* [Wanderings of Plato. The scholarly tradition during the Ottoman occupation], Αθήνα 1963.

Αγγέλου Α., "Το χρονικό της Αθωνιάδας" [The chronicle of the school at Mount Athos], *Νέα Εστία,* Χριστούγεννα, Αθήνα 1963.

Αποστολόπουλος Δ. Γ., "Για την προϊστορία του Νεοελληνικού Διαφωτισμού. Στοιχεία Φυσιολογίας τον 17ο αιώνα στην Κωνσταντινούπολη" [About the prehistory of the Neohellenic Enlightenment. Elements of physiology in Constantinople during the 17th century], *Ο ερανιστής,* 11, 1974, 296-310.

Γαβρόγλου Κ., "Οι επιστήμες στον Νεοελληνικό Διαφωτισμό και προβλήματα ερμηνείας τους" [The sciences during the Neohellenic Enlightenment and problems in their interpretation], *Νεύσις.* 3, 1995, 75-86.

Γεδεών Μ., *Η πνευματική κίνηση του Γένους κατά τον ιη' και ιθ' αιώνα* [The intellectual activities of the nation during the 18th and 19th centuries], editors Α. Αγγέλου - Φ. Ηλιού, Ερμής, Αθήνα, 1976.

Γεδεών Μ.. *Χρονικά της Πατριαρχικής Ακαδημίας* [The chronicle of the Patriarchal Academy], Constantinople 1883.

Γριτσόπουλος Τ. Α., Πατριαρχική Μεγάλη του Γένους Σχολή, [The Patriarchal Academy], 2 volumes, Αθήνα 1966.

Δημαράς Κ.Θ., *Νεοελληνικός Διαφωτισμός* [The Neohellenic Enlightenment], Ερμής. Αθήνα 1989.

Καράς Γ., *Φυσικές και θετικές επιστήμες στον ελληνικό 18ο αιώνα* [The physical and positive sciences during the Greek 18th century], Gutenberg, 1977.

Κονδύλης Π., *Ο Νεοελληνικός Διαφωτισμός - Οι φιλοσοφικές Ιδέες* [The Neohellenic Enlightenment –the philosophical ideas], Θεμέλιο, Αθήνα 1988.

Κριμπάς, Κ., "Ο Δαρβινισμός στην Ελλάδα. Τα πρώτα βήματα" [Darwinism in Greece. The first steps] , in *Θραύσματα Κατόπτρου,* Θεμέλιο, Αθήνα 1993, pp. 81-108.

Μπενάκης Λ.. "Από την ιστορία του μεταβυζαντινού αριστοτελισμού στον Ελληνικό χώρο. Αμφισβήτηση και υπεράσπιση του φιλοσόφου στον 18ο αι. Νικόλαος Ζερζούλης-Δωρόθεος Λέσβιος", [From the history of the post-byzantine aristotelianism in the Greek region. Doubting and defending the philosopher in the 18th century. Nikolaos Zerzoulis –Dorotheos Lesvios], *Φιλοσοφία,* 7, 1977, 416-454.

Νικολαΐδης Ε., Διαλέτης Δ., Αθανασιάδης Η., "Τυπολογία των βιβλίων των θετικών και φυσικών επιστημών του προεπαναστατικού αιώνα (1750-1821)" [Typology of the books of physical and positive sciences during 1750-1821], *Τετράδια Εργασίας*, 8, Κέντρο Νεοελληνικών Ερευνών, Αθήνα 1986.

Παπαδόπουλος Θανάσης, *Η νεοελληνική φιλοσοφία από τον 16ο έως τον 18ο αιώνα* [The Neohellenic philosophy from the 16th to the 18th century], I. Ζαχαρόπουλος, Αθήνα 1988.

Παπανούτσος Ε.Π., (editor) *Νεοελληνική Φιλοσοφία* [Neohellenic philosophy], Βασική Βιβλιοθήκη, Αθήνα, 2 volumes 1953, 1957.

Παρανίκας Ματθαίος, *Σχεδίασμα περί της εν τω ελληνικώ έθνει καταστάσεως των γραμμάτων από της αλώσεως της Κωνσταντινουπόλεως μέχρι της ενεστώσης εκαντοετηρίδος* [Outline of the situation concerning the state of letters from the fall of Constantinople to the present century], Constantinople 1867.

Σάθας Κ., *Νεοελληνική Φιλολογία. Βιογραφίαι των εν τοις γράμμασι διαλαμψάντων Ελλήνων ...(1453-1821)* [Neohellenic literature. Biographies of Greeks who have excelled in the letters 1453-1821], Αθήνα 1868.

Στεφανίδης Μ., *Αι φυσικαί επιστήμαι εν Ελλάδι προ της Επαναστάσεως. Η εκπαιδευτική επανάστασις* [The physical sciences in Greece before the revolution. The educational revolution], Αθήναι 1926.

AGUSTÍ NIETO-GALAN

THE IMAGES OF SCIENCE IN MODERN SPAIN

Rethinking The "Polémica"

"Spaniards have good qualities for Science, and many books. Nevertheless, perhaps Spain is the most ignorant nation in Europe. What can we expect of a country that needs to ask priests for permission to read and think?"

N. Masson de Morvilliers (1782)[1]

"As regards the subjects taught in secondary schools, someone once said that our defeat was inevitable, because the United States was a country of Physics and Chemistry, and Spain one of Rhetoric and Poetics"

J. Rodríguez Carracido (1898)[2]

"Spaniards, who tend to feel superior to the rest of the world in a series of virtues, in others - like science and technology - see themselves as inferior"

J.J. López Ibor (1951)[3]

INTRODUCTION

These quotations share a pessimistic view of scientific culture in Spain. They reflect a negative discourse according to which Spaniards are reluctant to assimilate science and technology, and indeed seem to lack the skills to do so. Although the core of the main debate dates from the second half of the nineteenth century, these texts are quite representative of multiple episodes throughout Spanish intellectual history, in which, for a variety of reasons, the country's scientific capacity was questioned in public discourses. The vision presented of the country's scientific talent and ability tended to be negative, although this pessimism was countered at times by an exaggerated apologetic and passionate defense of the nation's endeavours.

In this paper, I shall choose some of the most relevant episodes of that longstanding public controversy, commonly known today as "la polémica de la ciencia española".[4] I will try to analyze the effect that a frequent negative image of a 'weak' Spanish Science, and, as a result, the fervorous reaction of national proudness has had on the local historians of science and technology. Undoubtedly, any thorough examination of issues that reflect the main features of Spanish science requires a broad historiographic consensus about the 'real' achievements of the country. Thus, a review of the 'polémica' may contribute to a reassessment of some traditional historiographical problems, and to a fuller understanding of the role of science and technology and their public image in Spain.

Following up the "polémica", and tracing back some episodes of the controversy about the nature of Spanish science and its image among scientists, writers and intellectuals, this paper will analyze how a tacit and recurrent inferiority complex that the Spanish felt and expressed vis-à-vis Europe influenced the scientific debate itself, and as a result, shaped the way in which the history of Spanish science has been written and transmitted to younger generations. Because of this subtle feeling of belonging to a weak and peripheral scientific community, Spanish historians of science have often constructed their narratives stressing counter arguments, through a diffusionist model, accepting too easily a view of Spanish science as a mere imposition of a dominant scientific culture from the North, partially neglecting the study of the plurality of sites for creating and reproducing scientific knowledge. Some preliminary proposals for approaching the historical past of the country, trying to avoid the influence of that long-standing debate will be discussed in the last part of the paper.

1. THE HISTORY OF "BACKWARDNESS", AND THE DEBATE ON SUCCESS OR FAILURE

In 1949, the French historian Fernand Braudel defined three different categories of historical time in his famous book, *La Mediterranée et le monde méditerranéen à l'époque de Philippe II*. An individual time represented the frequent and accelerated particular human events; a social time was slower, it shaped deep economic and social changes; and, finally, a sort of geographic time, almost still or even cyclic was, in Braudel's view, a very useful tool to explain long-standing phenomena, the "long-durée".[5] Character of nations, cultural traditions, idiosyncratic social values might be explained through this long-standing historical analysis, which emphasizes continuity rather than change., and which might help to understand a complex historical problem as the Spanish "polémica".

During the reign of Philip II (1556-1598) astronomy, mathematics, shipping, artillery, medicine or botany were clearly promoted by the monarchy in order to provide useful tools for the organization and control of the Spanish Empire. In spite of the strong influence of the Catholic Church and the slow introduction of the new natural philosophy of the Scientific Revolution, there was no negative attitude among Spaniards against science and technology, which it seems did not play an important role in the late seventeenth century politic and economic decadence.[6] The

Spanish Empire and the downfall of the Habsburg dynasty, was mainly due to expensive religious wars, to the costly organization of a territory of enormous size and population, and to the feudal nature of the Spanish economy in a new emerging world of trade, manufacturing and protoindustrial investment.[7]

In the pessimistic atmosphere of the last decades of the seventeenth century, a dynamic circle of natural philosophers and physicians, the "novatores",[8] introduced and defended the new foreign concepts of the Scientific Revolution.[9] The "novatores" Juan Bautista Juanini, Juan de Cabriada,[10] following the iatrochemical ideas, challenged Gallenic explanations of breathing, animal heat, digestion, and the nature of blood. Thus, the *Carta Filosófica, Médico-Chymica*, published in Madrid, in 1687, by Juan de Cabriada,[11] firmly rejected Aristotelian authority. The "novatores" were also very keen to introduce experimental natural philosophy. They proposed the Baconian idea of progress to combat the dogma of scholastic authority. Intellectual backwardness could be overcome only through new links with Europe, and they distanced themselves from the old Spanish scientific tradition that dated from the Arabic and Christian Middle Ages[12] and the Habsburg projects of the Renaissance.[13]

The fact is that the spirit of the "novatores" continued in the first decades of the eighteenth century,[14] when, after the War of Succession, the French Bourbon dynasty took the Spanish Crown and began to plan political, economic and cultural reforms to 'modernize' the country, embodied principally in the reinforcement of its links with Europe.[15] The friar Benito Jerónimo Feijoo wrote in 1745 an essay on the causes of the Spanish backwardness in the natural sciences.[16] There were six main reasons to explain the lack of modern science in Spain: 1. the poor background of the teachers; 2. the reluctance to accept novelty; 3. the assumption that ideas of the new natural philosophers were useless curiosities; 4. the rejection of Descartes and the new philosophy; 5. the fears that the new philosophy could challenge scholasticism; 6. the tradition of emulation, copy, and lack of originality.

Fighting against that black picture, Charles III (1759-1788) launched the Bourbon plan for the modernization of public administration, economy and culture. It was a remarkable undertaking. Under Royal patronage, new scientific institutions were created, often inspired by the rhetoric of applied knowledge. Technical and medical schools, botanical gardens, "Sociedades económicas de amigos del país", trade boards ("Juntas de Comercio"), and new academies provided more flexible channels for the introduction of new sciences and technologies, in an intriguing alliance between the "Republic of Letters" and the French-influenced monarchy.[17] For many Spanish historians, the Enlightenment has become one of the 'mythical' periods of the history of Spanish science, a golden age to hark back to when discussing weaknesses and problems of research and education in later years.

Nevertheless, one of the most passionate episodes of the "polémica", and, for some historians, a sort of symbol of the starting point of this long-standing debate,[18] was the paper "Espagne", a famous article published in 1782 in the French *Encyclopédie Méthodique* by Nicolas Masson de Morvilliers. In the pre-revolutionary context of the "philosophes", Masson launched a fierce attack on the Spanish Ancien Régime and against the strong influence of religion, dismissing it as a country incapable of valid work in science and technology. Masson's paper has become one of the milestones of the "polémica". Incensed by its attacks, a number of prestigious Spanish writers[19] embarked on an impassioned defense of the country's scientific culture and tradition, and praised the Spanish resistance to the pernicious influence of the French "philosophes"; in contrast, others, like Luis Cañuelo in his papers in the journal *El Censor*,[20] supported Masson, denouncing Spain's backwardness, and resolutely defending the need for profound reforms and closer links with Europe, and with France in particular.

Nonetheless, the gap between rhetoric and reality seems to have been a wide one in late eighteenth century Spain. Some historians have argued that the Bourbon plans did not produce tangible results.[21] The impact of the French Revolution,[22] the hesitant policies of Charles IV, and the Napoleonic War led to deep political, economic and social tensions. The times were marked by the crisis of the Ancien Régime, caused in part by the conflict between the Catholic Church and the agrarian aristocracy on the one hand, and minority mercantile and industrial groups on the other.[23] The debate concerning Spanish science in the late eighteenth century was one of the numerous signs of conflict and tension in a society in rapid transformation. The new liberal ideas of the French "philosophes" and the Revolution[24] clashed with the Ancien Régime and the defenders of the monarchy; old universities such as Salamanca, Alcalá and Santiago with new alternative cultural institutions of the Enlightenment.

The early decades of the nineteenth century were characterized by tension and upheaval. The Napoleonic War lasted from 1808 to 1814; the absolutist régime of Ferdinand VII reigned from 1815 to 1820, and 1823 to 1833, punctuated by the Liberal triennium from 1821 to 1823. Most practitioners of natural philosophy and science were close to liberal ideas[25] or dissenting religious movements,[26] and many were engaged in the liberal plans of the Constitution of Cádiz in 1812, and in a range of educational projects designed to broaden scientific culture through French oriented programmes of instruction.

Unfortunately, not enough is known of the history of Spanish science in the mid-nineteenth century. Nonetheless, it seems clear enough that political tensions eased off to some extent, and a new generation of scientists, the "Isabelines",[27] emerged during the moderate liberal years of the reign of Isabel II, recovering the old Enlightenment idea of the importance of the scientific 'permeability' of Spain. The

"Academia de Ciencias Exactas, Físicas y Naturales", was founded in Madrid in 1847. Ten years later, in 1857, under a new educational policy, the Science Faculty (Physics-Mathematics, Chemistry, Natural Sciences) was created as a separate unit, though it was closely supervised by the state.[28] In the same period various Schools of Engineering were also founded.[29]

In spite of the moderate atmosphere of the Isabeline times,[30] and the intellectual freedom of the radical-liberal revolution of 1868,[31] the pessimistic image of Spanish science was at the core of public debates. Perhaps one of the most famous examples was the case of the engineer José Echegaray,[32] who expressed serious doubts about the capacity of Spaniards as mathematicians in a famous paper on the History of Pure Mathematics in 1866.[33] Against the backdrop of nineteenth century European nationalism, the arguments about the success or failure of the Spanish science fitted very well inside the discourse of Spain as a 'modern' nation-state, and it was not difficult to construct mythical origins for the "polemica".[34] After the revolutionary period of 1868-1874, the Bourbon Restoration provided a more stable political background for the promotion of science, but the theories emerging abroad at that time and seeking acceptance - such as Darwinism or psychoanalysis - were particularly controversial.[35] Significantly, the professor of 'Biological chemistry' at the University of Madrid, José Rodríguez Carracido,[36] defended the values of Spanish science during the Enlightenment. He was committed to the Spanish nation ("patria"), to liberal ideas, as well as to Darwinism.[37] Rodríguez Carracido gave numerous lectures: on the ideal conditions of Spain for the practice of science; on the problem of scientific research in Spain; and on the uses of chemistry. In 1908, he contributed to the diffusion of scientific values through the foundation of a Spanish Association for the Advancement of Science ("Asociación Española para el progreso de las Ciencias").[38]

The late nineteenth century debate focused mainly on the issues[39] of the utility of modern science[40] and the role of Europe in the development of Spanish science. The controversy about whether science was useful enough to claim public support had attracted considerable attention since the Enlightenment.[41] Juan Pablo Forner, a prominent defender of the virtues of Spanish science against Masson's thesis, was dismissive of 'speculative' sciences that had no applications to practices such as navigation, building, manufacture or warfare, etc.[42] In the late nineteenth century, the utility of science was again a matter of debate even among scientists and supporters of science. Thus, the leading Spanish intellectual Marcelino Menéndez Pelayo identified some of the causes of the Spanish backwardness, stressing the 'despotism of the utilitas', which put a brake on scientific creativity. In 1894, Menéndez presented his famous diagnosis of Spanish scientific illness: the lack of interest in the 'sublime utility of useless science'.[43] That is to say, a long-standing obsession for easily applicable science and technology contributed to marginalize pure and basic research.

The second great issue was Europe. The old eighteenth century "pensionados", who crossed the Pyrenees in search of new scientific theories and technical innovations and were deeply influenced by the French, brought back a pervasive sort of complex of inferiority, which was to last for centuries, and an obsession with imitating foreign achievements regardless of the quality of the local production. For the generation of the late nineteenth century, however, the reference point shifted from France to Germany; public speeches and lectures made dramatic calls for the reform of Spanish scientific research and teaching to copy the German style. Typical was the speech of Eugenio Mascareñas, professor of chemistry at the University of Barcelona, in 1899:[44]

> " To be aware of the regrettable situation of the experimental sciences in our country ... we should observe only two main disciplines, Physics and Chemistry, and compare their lack of promotion in Spain with the prosperous policies established by the governments of the most enlightened nations of Europe and America. At the head of this fruitful and civilizing movement...we should first cite Germany"

A third issue can be identified in the public discussion around 1900: the problem of religion, the 'black legend' of the Spanish Inquisition,[45] the political influence of the catholic Church, and its role in the history of Spanish science from the Renaissance onwards. It is no coincidence that at that time two very important books on the history of the relations between science and religion (Draper, 1873; White, 1876) were translated into Spanish. The local debate may have been influenced by White's idea of the lack of tolerance and free thinking in Spain in comparison with other European intellectual contexts.[46] Nevertheless, some historians believe that this 'black legend' of the Spanish Inquisition had been excessively influenced by the views of Anglo-Saxon historiography.[47]

After the loss of two very important Spanish colonies, Cuba and the Philippines, in 1898 - a loss which symbolized the last days of the glorious Spanish Empire - the whole country engaged in intense self-criticism, in search of the causes of the terrible military defeat. The nation's scientific backwardness was often quoted as a decisive reason.[48] Santiago Ramón y Cajal, winner of the Nobel Prize for Medicine (1906), insisted on the crucial importance of the scientific 'Europeanization' of Spain in his public speeches,[49] and the philosopher José Ortega y Gasset warned of the dramatic inferiority of Spain in relation to Europe, and of the irrelevance of Spanish science.[50] This idea was closely linked to the movement of 'regeneration' of the country inside a dynamic debate on the causes of the loss of Spanish influence in the international arena.[51]

Ramón y Cajal agreed that Spain was indeed backward, but, by opposition to Menéndez Pelayo's views, he argued that the real causes lay - as had been claimed at the end of the late eighteenth century - in educational policies, and not in ethnic, geographical, moral or religious issues. To gain a clearer understanding of how far the "polémica" went, it is interesting to see the long list of essentialist theories that were proposed as explanations of the weakness of the scientific culture in Spain, which Ramón y Cajal systematically refuted in his papers:[52]

1. Physical theories:
   a. Thermal hypothesis (Hot weather);
   b. Oligohydrid theory (Lack of rain);
2. Moral and political theories: economic, political and military weakness;
3. Religious theories: the Inquisition and religious fanaticism;[53]
4. Spanish pride and arrogance (aristocratic distaste for mechanical work, trade and industry);
5. The tradition of intellectual isolation of Spain.

These theories were doubtless exaggerated, and at the turn of the century, contingent factors, like the improvement of education and the institutional support for science and technology became matters for concern in many intellectual circles. In spite of the weakness of the University and the limitations of the official plans for the promotion of science during the Bourbon Restoration (1875-1923), a new 'civil discourse'[54] emerged in the context of reform projects and intellectual debates about the causes of the country's political decadence. Different lobbies tried to find social agreements to ease political tensions, to promote pure science, and to create a more pragmatic atmosphere, especially after the defeat of 1898.[55]

In 1907, Ramón y Cajal was made president of the "Junta para Ampliación de Estudios", an institution devoted to the promotion of education and research. It continued the old tradition of the eighteenth-century "pensionados" and the enduring desire to open the country up to Europe. It also had also close links with the free thinking movement (the Krausists),[56] which regarded Spain as an immense classroom. The ideas of freedom, rationalism, honesty and admiration for science were to be disseminated, and only closer contact with Europe could combat dogmatism, irrational superstition, and religious hypocrisy.[57]

As far as the debate on the utility of science was concerned, the fact is that Engineering Schools were quite active, particularly in industrial areas such as Barcelona. In this regard, the Catalan astronomer Josep Comas Solà published an article in the newspaper *La Vanguardia*, in 1899, entitled: "Nuestra decadencia" (Our decadence).[58] Comas denounced the lack of a scientific policy and the useless investments in second-rate literary education, or in the building of new bull-rings, and the absence of papers written by Spanish scientists in the international journals full of studies by 'European' researchers. The reasons that he gave for that isolation were:[59]

"....in Spain, there is a dramatic lack of education (12 million of people are not able to read), and among the few scientists, the spirit of research and study and the love of work is almost zero...there is a lack of theoretical and practical science, love of progress, responsibility and spirit of improvement".

These passionate controversies and debates in favour or against the existence of a real scientific culture in Spain continued during the twentieth century. In the 1950s the historian of medicine Pedro Laín Entralgo,[60] in an attempt to describe the long-standing bipolar tension in the Spanish culture, proposed a set of dialectic definitions:[61] rebellion vs. obedience; efficiency vs. utopianism; unversalism vs. relativism; empiricism vs. formalism; lack of interest vs. search for profit; improvisation vs. method. The myth of the 'two Spains' also included two ways of understanding science, partially represented, in Laín's view, by Menéndez Pelayo and Ramon y Cajal; he insisted on the importance of the history of science as a key issue for any rigorous debate.

In fact, the debate on the causes of the Spanish scientific backwardness and the discussion about the different images of science continued during the Civil War (1936-1939), when the myth of the 'two Spains' that previous generations had constructed finally reached its climax in the form of social and political conflict. In the 1940s, faced with the emergence of a military regime, a substantial part of the scientific and intellectual community went into exile. In spite of this, some of the leaders of the Spanish scientific policy under the Franco regime made selective use of examples from the past in their bid to legitimize the targets of the new National Research Centre ("Consejo Superior de Investigaciones Científicas" CSIC).

In 1949, the chemist Manuel Lora Tamayo[62] praised Menéndez Pelayo, describing him as a remarkable example in the fight against scientific pessimism and in the struggle to promote the famous concept of the 'utility of useless science' in the new University and research system designed by the regime. In 1950, the Secretary of the CSIC, J.M. Albareda,[63] criticized the pessimism of papers and books from the late nineteenth and early twentieth century concerning Spanish scientific skills for invention and research.[64] In Braudel's terms of the "long durée", almost 200 years after Masson's article in the *Encyclopédie Méthodique*, the negative feelings concerning Spanish science persist, and even prominent figures of the democratic period - after Franco's military dictatorship - complained about the difficulties facing scientific research and education in the Spanish context. In 1985, Ramon Margalef, an outstanding professor of Ecology at the University of Barcelona, wrote:[65]

"I regret to say that the Spanish scientific arena has not changed. This is not a problemof organization but of lack of ideas. I do not see any movement which allows us to envisage that we are on the threshold of the renaissance of Spanish science"

2. THE IMAGES OF SCIENCE AND ITS INFLUENCE ON THE SPANISH HISTORIANS

There is no clear agreement among historians about the real impact of these public controversies in Spanish cultural circles on the actual practice of science and technology, and on whether that nineteenth-century debate, with even older historical references like the "novatores" or Masson, properly reflects an objective, historical vision of Spanish science. Nonetheless, there is no doubt that the problem of understanding the scientific past was always in the heart of the discussion, and played a key role in the construction of the images of science from the eighteenth century onwards. Obviously, it was necessary to find convincing historical examples to show the virtues or the failures of Spanish science and technology, and in this regard, all those debates, the history and the historiography of Spanish science have multiple connections, which might shed some light on the understanding of the particular conditions of that local context. Thus, one could argue that the intellectual public controversies had a certain influence on the ways in which Spanish historians of science saw their own past. They may also explain the existence of a frequent 'tacit agenda' which stressed the major achievements of the Spanish scientific past.[66]

In the nineteenth century, the history of science was mainly written by scientists or by educated people interested in the scientific culture of the country and in the broad debate about the 'real' contribution of Spain to the Mankind. Historical examples were often used to promote new scientific policies, to encourage new research projects, or even to reinforce liberal, patriotic ideas.[67] Numerous scientists wrote books on the history of their own disciplines, using this research to call attention to the aspects of the practice of science in need of support from public or private institutions. Rodríguez Carracido followed this line in his critical-historical approach to Spanish science.[68] He wrote a history of chemistry,[69] and also the history of important eighteenth century scientists and their connections with the Spanish Enlightenment.[70] Mascareñas gave lectures to popularize chemistry in the Academy of Barcelona and provided a historical reconstruction of the chairs of chemistry in the Spanish universities during the nineteenth century.[71] In 1934, Enric Moles, one of a number of 'chemist-historians',[72] gave a lecture on 'El Momento científico español, 1775-1825',[73] in which he spoke of the generation of the end of the eighteenth century, a generation that institutionalized the chemistry that was emerging from the Lavoisierian revolution, and introduced Spanish audiences to the new discipline.[74]

In order to find convincing arguments for an optimistic vision of science, an extremely erudite history of Spanish scientists was developed. In the eighteenth century, a sort of 'protohistoriography' of science was introduced, for example in the

Catalogues of Spanish naturalists.[75] Between 1842 and 1852, A. Hernández Morejón published a bibliographical history of Spanish medicine,[76] and some years later, Menéndez Pelayo wrote *La Ciencia Española* (first issued in 1876 and reedited many times). The third volume of this work compiled all Spanish authors and books published in all cultural domains from the Middle Ages to the eighteenth century,[77] together with his *Inventario Bibliográfico de la ciencia española*.

In fact, one of the main arguments of the critics against Spanish science was the serious lack of great luminaries. In the late nineteenth century, exhaustive lists of foreign figures were published to show the striking absence of Spanish names in the top ranks of international science.[78] Scholarly histories of Spanish sciences, propsopographies, bibliographies, catalogues of scientific books became a genuine Spanish genre, which persists even today, in particular in the works of the Valencia group founded by Professor José María López Piñero, the "Instituto de Estudios Documentales e Históricos sobre la Ciencia".[79]

In his work on early modern Spanish science, López Piñero[80] criticized the focus on the search for great figures, wisely defending the need of a rigorous empirical research. In his view, once new generations of Spanish historians of science had recovered data about the scientific past of the country, the bitter controversies concerning the success or failure of Spanish science would become a sterile, minority debate, unrepresentative of the historical past. His book on the science in the sixteenth and seventeenth century Spain was an excellent pioneering example of the new historiography.[81]

As the historian of Medieval Science J.M. Millás Vallicrosa[82] claimed in 1956,[83] from the Enlightenment onwards, Europe emerged as a crucial reference point for the achievements of Spanish science. At the end of the nineteenth century, the Krausists, the free-thinking lay movement, suggested that because of the failures of the State and the control of the Inquisition, Spanish science actually did not exist. This radical assumption dismissed the Spanish scientific past, and was strongly criticized by Menéndez Pelayo, who emphasized not only the erudite collection of Spanish names and books, but the need for rigorous research into the history of Medieval and Renaissance science in Spain, including the rich Arabic past, the botanic expeditions to the New World or the imperial science of the sixteenth and seventeenth centuries in domains such as cosmography, cartography and shipbuilding.[84]

In spite of some historical inaccuracies in Menéndez's work, Millás acknowledged his great contribution to the defense of the Spanish scientific and philosophical tradition, and his emphasis on the great importance of Arabic and Jewish works in the Middle Ages, the scientific translations of the School of Toledo, the mystic theology and technology of the sixteenth century, and the eighteenth century botany and metallurgy in the South American colonies.[85] Following Menéndez's suggestions, Millás made important contributions to the history of Medieval Spanish science, later continued by two outstanding historians, Joan Vernet and his pupil Julio Samsó.[86]

In addition, the history of Spanish science in the Enlightenment, and particularly during the reign of Charles III (1759-1789) has attracted wide-ranging research.[87] Even in the democratic period since 1975, historians and in particular historians of science have tried to recover some of the ideas and policies of the Enlightenment, in an attempt to link Spain with the cultural progress of Europe since the end of the Second World War. In fact, the 'obsession' with Europe, like the debate about the nature of a genuine Spanish scientific culture has been a long-standing feature of the discussions of the different generations: "novatores", "pensionados", "liberalas", "regenerationistas", and "krausistas".

These debates have stimulated research into the history of Spanish science, a fact borne out by the scholarly interest in the work of the nineteenth-century scientist-historians. However, the emphasis on local figures in their local context with little concern for the international arena might be a less palatable historiographical consequence of those debates. In spite of the rhetorical proclamations of the urgent need to avoid sterile debates about the virtues and vices of Spanish science, even López Piñero - whose intellectual authority is widely acknowledged - has come in for criticism for his assessment of the "polémica" as an exaggeration, or an excuse for other political, social, economic or ideological tensions, and for his attempts to wrest attention away from it in his *Ciencia y Técnica en la sociedad española de los siglos XVI y XVII*. For example, for Antonio Márquez, the figure of Masson still influences López Piñero's work.[88] Indeed, it is not easy to escape the use of "Europe" as a permanent reference point, and according to López Piñero this has pushed Spanish and other scholars in recent decades towards an intellectual dependence on the British and American historiography.[89]

Spanish science and technology have been heavily dependent on foreign achievements. This is a fact that histories of science cannot avoid. Recent Spanish papers frequently refer to concepts such as "the reception of modern chemistry", "the reception of Darwinism" "the reception of relativity", or pay particular attention to the names of Spanish scientists who were pupils, students, friends or followers of great European figures - Paracelsus, Descartes, Linneus, Cuvier, Lavoisier, Darwin, Einstein. They also frequently show great interest in measuring the time gap between a particular European invention of a new machine or theory and the moment of its reception, or imitation, in Spain.[90]

'Second class science' (as an uncharitable critic might describe the contributions of the Spanish) has often been written using a narrow historiographical approach. López Piñero argues against the mechanical imitation of recent British and American historiography, in his bid to recover other German, French or Italian traditions in the history of science of the early 20th century. However, the problem often lies in the difficulties in defining more useful questions for research, which would avoid all the pitfalls of terms like 'success', 'failure', 'better or worse', 'copy or invention', 'reception', or 'periphery'.

Other historians who have focused their research on the history of Spanish industry, economy and technology have emphasized the idea that a scientific culture should often be linked with a dynamic industrial milieu. This was another key factor in understanding why Spain, a latecomer to the Industrial Revolution, failed scientifically.[91] For these historians, the elitist intellectual and academic discussions about the qualities of Spanish science and culture are irrelevant; however, the low quality of Spanish technology is tacitly assumed as a key factor in the slow rhythm of the country's industrialization.[92]

Perhaps one explanation for the minority position of the history of science at institutional level in Spain lies in this long-term, tacit pessimism. Why study the scientific past of a country considered insignificant as far as science and technology are concerned? The institutional research and teaching in the history of art in Spain stands out starkly compared with the general lack of interest shown in science. Similarly, the history of medicine has attracted attention in Spanish university circles, conceivably due to the clearer social connections of these disciplines. New historiographical trends in the social history of medicine seem more integrated in the research methods of the numerous Spanish historians active in the discipline.

In spite of the efforts of the many historians of Spanish science who have striven to play down the influence of the "polémica", the fact is that some of the distorted views borrowed from it are still visible in modern times. The following paragraph, written in 1979 by López Piñero, discusses the nature of Spanish science in the nineteenth century:[93]

> "[Spanish science] depends on the endeavours of one man, or of a specific group of men who manage to establish European contacts, and in some isolated cases, take part in important international scientific contributions, but they work against a background of complete indifference inside their own society".

3. TOWARDS A RENEWED HISTORY OF THE SPANISH SCIENCE

Obviously, there is no easy solution to a complex problem such as this. Probably the most difficult question is to explain the importance of the history of science in a country which on the whole has paid scant attention to its history - let alone the history of science, a subject considered a second-rate activity by many generations of Spaniards.

Of course, much more research is needed to generate a new historiography which could eventually challenge hegemonic interpretations on sensitive issues such as 'science and religion' - for example, the role of Catholicism and Inquisition. Other areas that require analysis are the social impact of technology - for example, the advent of the new domestic appliances in the 1960s - and the teaching of science - or the nineteenth-century state plans for the establishment of a scientific syllabus.

Perhaps one way to avoid a narrative often trapped in the "polémica" would be to introduce new approaches that have emerged over recent years under the titles of "the culture of science and technology", "the public understanding of science", and "science and the public", etc.[94] In these views, science and technology are much more than the achievements of the great luminaries. They refer to a set of cultural beliefs, values and practices in different parts of the society. Science can be seen as a social strategy to spread particular values, to contribute to the establishment of social order, to create a new language of communication in a network of practitioners. Science also concerns technological devices that have always been closely linked to the everyday life of men and women. Science is universal, but also local and particular. In the same way as cultural anthropologists emphasize different values in different cultures,[95] scientific practices in particular contexts are not easy to extrapolated to others.

The tacit assumption of the universality of science is a source of a certain uneasiness among historians of Spanish science in their attempts to understand the specific role of science and technology in this country. The use of windmills, or astrolabes in the Middle Ages, the botanical expeditions to the New World, the image of science during the times of the Inquisition, the culture of artisans, the history of public health, science teaching in secondary schools, and many other research subjects can be of great interest to a wide audience in Spain and abroad. These historical problems should be presented with a rich collection of new empirical data, but combined to a new way of seeing science in its cultural context,[96] with a new set of research questions to avoid sterile controversies.

By reconstructing the epistemologies of popular scientific and technological experiences[97] in local settings in Spain, a new framework of historical interpretation of the numerous and very rich historical sources can be created. The study of the history of Spanish science via the role of specific practices (lectures, experiments, informal gatherings), machines and instruments would probably be more fruitful than a too erudite approach to the Spanish translations of the most important works of the great foreign 'savants'. How the cultures and practices of science and technology in Spain were integrated in the everyday life of a group of individuals in a particular historical context is a question of considerable relevance.

On the other hand, there is no doubt that the recurrent intellectual debates about Spanish science, particularly the nineteenth-century vision of the "polémica", were framed in an era of the rise of European nationalism, in which the most powerful countries were engaged in fierce competition to convince their foreign rivals of the excellence of their own culture and civilization. For example, the image of French science after the Franco-Prussian War, or the links between science and technology in Germany during the First World War reflect the common endeavour to construct a national image for French, German or Spanish science, a desire that was quite habitual from the late nineteenth century onwards; in addition, there was a considerable gap between the rhetoric and the reality of science in many countries, not only in Spain. In this sense, the nineteenth-century "polémica" was part of a broader reflection on the public images of national and foreign science in western societies,[98] with, perhaps only one genuinely Spanish ingredient, the strong emphasis on the pessimistic view of the country's backwardness in relation to Northern Europe.

New trends in the sociology of science and technology[99] have placed special emphasis on the importance of 'symmetrical history', able to study particular accepted and ignored experiments, theories, inventions or machines. This symmetry would naturally help to avoid the obsession with the great names, the foreign 'celebrities' who appear in the introductions of Spanish secondary school textbooks in the study of any sort of scientific and technical endeavours. A symmetrical, alternative method to explain the history of Spanish inventions and innovations would also be of great interest. Moreover, the ways in which science and technology shape society might provide another powerful method to understand the meaning of particular scientific practices in local contexts such as Spain.[100]

The role of science has been extensively explored in other cultural contexts such as Victorian Britain. Perhaps I cannot avoid the influence of the idea of the "Europeanization" of Spanish research, which has been often at the heart of the "polémica" of the Spanish science, but masterpieces such as Morrell and Thackray's *Gentlemen of Science* are immensely stimulating for any reader keen to understand the role of science in nineteenth century Britain. In that nineteenth century context science was:[101] "a form of rational amusement, theological edification, polite accomplishment, technological agent, social anodyne and intellectual ratifier of the new industrial order".

Of course, this deep immersion in a particular milieu in which science and technology are at the core of the historical narrative cannot automatically be applied in the Spanish context, but a method of this kind might conceivably lead to a better understanding of the scientific culture which has emerged in Spain in a range of periods, and to avoid discussions about great geniuses, ethnic aspects, pessimism or aggressive optimism. Perhaps at the end of the nineteenth century, one of the key moments of the construction of the "polémica", Spanish science was perceived as a potentially useful tool for the process of industrialization; a contribution to the implantation of laicism; the core of the educational plans of a new liberal state; an exotic practice of foreigners; a dream of modernization. I wonder if twenty-first century historiography will confirm some of these bold hypotheses, which have no winners or losers, and no first class and second class science.[102]

Universitat Autònoma de Barcelona

NOTES

[1] "El español tiene aptitud para les ciencias, existen muchos libros y, sin embargo, quizás sea España la nación más ignorante de Europa. ¿ Qué se puede esperar de un pueblo que necesita permiso de un fraile para leer y pensar?" MASSON DE MORVILLIERS, N. "Espagne" in "Géographie Moderne," Vol I, pp. 554-68. *Encyclopédie Méthodique.* Paris 1782. Quoted from the Spanish translation: GARCIA CAMARERO, Ernesto y Enrique (eds.). *La Polémica de la ciencia española.* Alianza. Madrid 1970, p. 51.

[2] " Refiriéndose a los títulos de las asignaturas de la segunda enseñanaza, alguien dijo donosamente que nuestra derrota [Cuba y Filipinas, 1898] era inevitable, por ser los Estados Unidos el pueblo de la Física y la Química, y España el de la Retórica y Poética". Cited by SANCHEZ RON, J. M. (ed.) *Ciencia y Sociedad en España.* CSIC. Madrid 1988. p. 14 (Introduction).

[3] "El español, que suele considerarse superior al mundo en una larga serie de virtudes, en otras se considera inferior. Sobre alguna de ellas - la ciencia, la técnica, - ha montado su complejo de inferioridad del español". LOPEZ IBOR, J. J. *El español y su complejo de inferioridad.* Rialp. Madrid 1951. (publicity leaflet).

[4] Numerous papers and books trace the history of the age-old debate about the 'virtues' and 'vices' of the scientific culture of Spain. A good starting point is: LOPEZ PIÑERO, J.M. *Ciencia y Técnica en la sociedad española de los siglos XVI y XVII.* Labor. Barcelona 1979. (Introduction). GARCIA CAMARERO, Ernesto y Enrique (eds.). *La Polémica de la ciencia española.* op. cit. CASTRODEZA, C. "A vueltas con la historia de la ciencia española: el problema de la idiosincrasia," *Sylva Clius,* 6, 1988, 299-330.

[5] " Ce livre se divise en trois parties, chacune étant en soi un essai d'explication. La prémier met en cause una histoire quasi immobile, celle de l'homme dans ces rapports avec le milieu qui l'entoure; une histoire lente à couler et à se transformer, faite bien souvent de retours insistants, de cycles sans fin recommencés" BRAUDEL, F. *La Mediteranée et le monde méditerranéen à l'époque de Phillippe II.* A. Colin. Paris 1949. pp. xiii-xiv.

[6] LOPEZ PIÑERO, J.M., *Ciencia y Técnica en la sociedad española de los siglos XVI y XVII. op. cit..* VICENTE MAROTO, M.I., ESTEBAN PIÑERO, M. *Aspectos de la ciencia aplicada en la España del siglo de oro.* Junta de Castilla y León. Salamanca 1991. GARCIA-TAPIA, N. *Patentes de invención españolas en el siglo de oro.* Ministerio de Industria y Energía. Madrid 1990. GOODMAN, D.: *Power and Penury. Governement, Technology and Society in Phillip II Spain.* CUP, Cambridge 1988. (Spanish translation 1990). GOODMAN, D. "The Scientific Revolution in Spain and Portugal" in PORTER, R., TEICH, M. (ed.) *The Scientific Revolution in National Context.* CUP. Cambridge 1992, 158-177.

[7] A vast historiography offers different explanations for the decline of the Spanish Empire in the seventeenth century: KAGAN, R.L. "Prescott's Paradigm: American Historical Scholarship and the Decline of Spain," *The American Historical Review*, 101, 1996, 423-446. KAMEN, H. *Spain in the later seventeenth century, 1665-1700*. Longman. London 1980. ELLIOT, J. H. *Spain and its World 1500-1700: Selected Essays*. Yale University Press, New Haven, 1989.

[8] LOPEZ PIÑERO, J.M. *Ciencia y Técnica en la sociedad española de los siglos XVI y XVII*. op. cit. MARTINEZ VIDAL, A., PARDO TOMAS, J. "In tenebris adhuc versantes. La respuesta de los novatores españoles a la invectiva de Pierre Régis," *Dynamis*, 15, 1995, 301-340. MARTINEZ VIDAL, A., PARDO TOMAS, J. "El Tribunal del Protomedicato y los médicos reales (1665-1724): entre la gracia real y la carrera profesional," *Dynamis*, 16, 1996, 59-89.

[9] On heliocentric cosmology and the introduction of some of the main ideas of the Scientific Revolution in Spain, see, for example: NAVARRO BROTONS, V. "The Reception of Copernicus in Sixteenth-Century Spain" *Isis*, 86, 1995, 52-78. ESTEBAN PIÑEIRO, M., GOMEZ CRESPO, F. "La primera versión castellana de De revolutionibus orbium Caelestium: Juan Cedillo Díaz (1620-1625)," *Asclepio*, 43, 131-162.

[10] LOPEZ PIÑERO, J.M. "Juan de Cabriada y el movimiento novator de finales del siglo XVII. Reconsideración después de treinta años," *Asclepio*, 45/1, 1993, 3-53.

[11] CABRIADA, J. *Carta philosophica, medicochymica en que se demuestra que de los tiempos y experiencias se han aprendido los mejores remedios contra las las enfermedades*. L.A. de Bedmar. Madrid 1687.

[12] SAMSO, J. *El legado científico andalusi*. Madrid 1994. VERNET, J. *Historia de la Ciencia española*. Instuto de España. Cátedra "Alfonso X el Sabio". Madrid 1975. SAMSO, J. *Islamic Astronomy and Medieval Spain*. Variorum. Aldershot 1994.

[13] This is, at least, López Piñero's thesis in his *Ciencia y Técnica...* op. cit. pp. 16-17.

[14] For the continuity of the ideas of the "novatores" in the early decades of the eighteenth century see: MARTINEZ, A. "Los orígenes del mito de Oliva Sabuco en los albores de la Ilustración" *Al Basit*, 13/22, 1987, 137-151. For a quite well known reflection of the Spanish Science in the mid eighteenth century see: FEIJOO, B.J. "Causas del atraso que se padece en España en orden a las ciencias naturales". *Cartas*, Vol II, 1745. In GARCIA CAMARERO, Ernesto y Enrique (eds.). *La Polémica de la ciencia española*. op. cit. pp. 25-43.

[15] For the historical evolution of the idea of modernization in Spain see: MARAVALL, J.M. *Antiguos y Modernos*. Alianza. Madrid 1986. MARAVALL, J.M. *Estudios de la Historia del pensamiento español del siglo XVIII*. Mondadori. Madrid 1991.

[16] FEIJOO, B.J., "Causas del atraso que se padece en España en orden a las ciencias naturales" *Cartas*, Vol II, 1745, in GARCIA CAMARERO, Ernesto y Enrique (eds.) *La polémica de la ciencia española*. op. cit. pp. 25-43.

[17] LAFUENTE, A. SELLES, M., PESET, J. L. *Carlos III y la Ciencia de la Ilustración*. Alianza. Madrid 1988. SANCHEZ-BLANCO PARODY, F. *Europa y el pensamiento español del siglo XVIII*. Alianza. Madrid 1991. ALVAREZ DE MIRANDA, P. *Palabras e ideas. El léxico de la Ilustración temprana en España (1680-1760)*. Anejos del Boletín de la Real Academia Española. Madrid 1992. MESTRE, A. *Corrientes interpretativas actuales de la Ilustración española*. In *España a finales del siglo XVIII*. Ediciones de la Hemeroteca de Tarragona. Tarragona 1982.

[18] GARCIA CAMARERO, Ernesto y Enrique (eds.) *La polémica de la ciencia española*. op. cit. pp. 8-9.

[19] CAVANILLES, A.J. *Observations de M. l'Abbé Cavanilles sur l'article Espagne de la Nouvelle Encyclopédie*. Jombert. París 1784. (A Spanish translation issued in the same year,

1784). DENINA, C. *Respuesta a la pregunta: ¿Qué se debe a España?* A speech to the Academy of Berlin (26-01-1786). FORNER, J.P. *Oración apologética por la España y su mérito literario.* Madrid. Imprenta del Real 1786.

[20] CAÑUELO, L. "Contra nuestros apologistas" *El Censor,* "Discurso CXIII," VI, 1786, 841-68. CAÑUELO, L. "La congoja de no poder hacerme entender de aquellos bárbaros", *El Censor,* "Discurso CLXI," VIII, 1786, 565-79.

[21] PUERTO SARMIENTO, F.J. *La ilusión quebrada. Botánica, sanidad y política científica en la España Ilustrada.* CSIC. Madrid 1988. GOODMAN, "Science and the Clergy in the Spanish Enlightenment" *History of Science,* 21, 1983, 111-141. DOMINGUEZ ORTIZ, A. *Sociedad y estado en el siglo XVIII español.* Ariel. Barcelona. 1976. p. 494.

[22] See HERR, R. *España y la Revolución del siglo XVIII.* Aguilar. Madrid 1988. (1st. English ed., Princeton, 1960). Chapter VIII: "El Pánico de Floridablanca".

[23] FONTANA, J. *La cirsis del antiguo régimen (1808-1833).* Crítica. Barcelona 1979.

[24] ARTOLA, M. *Los afrancesados.* Alianza. Madrid 1989. ARTOLA, M. *Antiguo Régimen y revolución liberal.* Ariel. Barcelona, 1979.

[25] PESET, J.L., GARMA, S., PEREZ GARZON, J.S. *Ciencias y enseñanza en la revolución burguesa.* Siglo XXI. Madrid 1978.

[26] GOODMAN, D. "Science and the Clergy...op. cit.

[27] VERNET, J. *Historia de la Ciencia española.* op. cit. J.M. LÓPEZ PIÑERO (ed.) *La ciencia en la España del siglo XIX. Ayer,* 7 (Madrid: Marcial Pons, 1992).

[28] MORENO GONZALEZ, A. "De la física como medio a la física como fin. Un episodio entre la Ilustración y la crisis del 98" in SANCHEZ-RON, J.M. (ed.), *Ciencia y Sociedad* ...op. cit. pp. 27-70, p. 67. GLICK, Th. *Einstein in Spain.* PUP. Princeton 1988. pp. 3-16. J.M. LÓPEZ PIÑERO (ed.) *La ciencia en la España del siglo XIX.* op. cit.

[29] RIERA, S. "Industrialization and Technical education in Spain, 1850-1914" in FOX, R. GUAGNINI, A. (eds.) *Education, technology and industrial performance in Europe, 1850-1939.* CUP. Cambridge 1993. 141-170.

[30] The Spanish Society of Natural History, created in 1871, had an active group of naturalists, who sought a more balanced position in between "unconscious pride" and "sad pessimism". LOPEZ OCON, L. "Ciencia e historia de la Ciencia en el sexenio democrático. La formación de una tercera via en la polémica de la ciencia española," *Dynamis,* 12 (1992), 87-104. pp. 91-92.

[31] LOPEZ OCON, L. "Ciencia e historia de la Ciencia..." op. cit.

[32] SANCHEZ-RON, J.M. (ed.) *José Echegaray.* Fundación Banco Exterior. Madrid 1990.

[33] ECHEGARAY, J. "Historia de las Matemáticas Puras en nuestra España". Discurso de entrada en la Real Academia de Ciencias Exactas, Físicas y Naturales. Aguado. Madrid 1866.

[34] SCHROEDER-GUDEHUS, B. "Nationalism and internationalism" in OLBY, R.C., CANTOR, G.N., CHRISTIE, J.R.R., HODGE, M.J.S. (eds.) *Companion of the History of Modern Science.* Routledge. London 1990. pp. 909-919. PAUL, H.W. *The sorcerer's aprentice. The French scientist's image of German Science, 1840-1919.* Gainesville, 1972.

[35] GLICK, Th. (ed.) *The comparative reception of Darwinism.* University of Texas. Austin 1974. "Spain," pp. 307-345. J. SALA CATALA, "Ciencia Biológica y polémica de la Ciencia en la España de la Restauración" in SANCHEZ-RON, J.M. op. cit. pp. 157-177. GLICK, Th. "El impacto del psicoanálisis en la psiquiatría española de entreguerras" in SANCHEZ-RON, J.M. op. cit. pp. 205-223.

[36] RODRIGUEZ CARRACIDO, J. *Estudios histórico-críticos de la ciencia española*. Alta Fulla. Barcelona 1988. (Introduction by A. Moreno González and Jaume Josa).

[37] RODRIGUEZ CARRACIDO, J. *Estudios histórico-críticos*..op. cit. (Introduction pp. vi-lviii).

[38] RODRIGUEZ CARRACIDO, J. *Estudios histórico-críticos*..op. cit. (Introduction p. xli).

[39] LOPEZ PIÑERO, J.M. *Ciencia y Técnica*...op. cit. p.24.

[40] MORENO GONZALEZ, A. *Una ciencia en cuarentena. Sobre la física en la Universidad y otras instituciones académicas desde la Ilustración hasta la crisis finisecular del XIX*. CSIC. Madrid 1988. In particular "El ser o no ser de la Ciencia Española: De la polémica sobre la utilidad de las Ciencias útiles a la polémica sobre 'la sublime utilidad de la ciencia inútil" pp. 438-445.

[41] MARAVALL, J. A. "El principio de la utilidad como límite de la investigación científica en el pensamiento ilustrado" a *Estudios de la Historia del pensamiento español del siglo XVIII*. op. cit. 476-488.

[42] MORENO GONZALEZ, A. *Una ciencia en cuarentena*. op. cit. p. 441.

[43] MORENO GONZALEZ, A. *Una ciencia en cuarentena*. op. cit. p. 444.

[44] "Para convencernos del lamentable estado de las ciencias experimentales en nuestro...basta, a mi juicio, que la atención se fije en dos de las más importntes, la Física y la Química, y compare lo que en España se ha dejado de hacer, con la pródiga y maternal solicitud que hacia ellas mostraron siempre, en este siglo los gobiernos de todas las naciones ilustradas de Europa y América. A la cabeza de este movimiento civilizador...es de justicia histórica citr en primer término a Alemania" MASCAREÑAS, E. *Consideraciones generales acerca de la ensañanza y estudio particular del estado en que se halla la de las Ciencias Experimentales en España*. Discurso Inaugural. Universidad de Barcelona 1899. 39-69, pp. 58-59.

[45] PEROJÒ, J., "La ciencia española bajo la Inquisición" *Revista Contemporánea*, 15-IV-1877. GARCIA CARCEL, R. *Orígenes de la Inquisición española. El Tribunal de Valencia, 1478-1530*. Peninsula. Barcelona 1976.

[46] MORON ARROYO, C. "Ciencia, Inquisición, ideología. Temas de nuestro tiempo," *Arbor*, 23, 1986, 29-44. p. 41.

[47] This could be the case of Henry Kamen, cited in MARQUEZ, A. (ed.) *Ciencia e Inquisición. Arbor*, 24, 1986. (Introduction, p. 9). This is a very interesting part of a more general debate on science and religion, which in the Spanish case could be traced back to: GIL NOVALES, A. "Inquisición y ciencia en el siglo XIX," *Arbor*, 23, 1986, 147-170. MARQUEZ, A. (ed.) *Ciencia e Inquisición. Arbor*, 24, 1986. MARQUEZ, A. *Literatura e Inquisición en España 1478-1834*. Taurus. Madrid 1980. PARDO TOMAS, J. *Ciencia y censura. La Inquisición española y los libros científicos en los siglos XVI y XVII*. CSIC. Madrid 1991.

[48] RODRIGUEZ CARRACIDO, J. *Estudios histórico-críticos*...op. cit. (Introduction, p. xxix).

[49] RAMON Y CAJAL, S. "Deberes del Estado en relación con la producción científica" Speech at the Academy of Science. Madrid, 5 December 1897.

[50] LOPEZ PIÑERO, J.M. *Ciencia y Técnica*...op. cit. p.24.

[51] RAMON Y CAJAL, S. "Deberes del Estado en relación con la producción científica" op. cit.

[52] RAMON Y CAJAL, S. "Deberes del Estado en relación..." op. cit. pp. 377-395. Cited in GARCIA CAMARERO, E.y E. (eds.). *La Polémica de la ciencia española*. op. cit.

[53] MARQUEZ, A. (ed.) *Ciencia e Inquisición*. op. cit.

[54] GILICK, Th. *Einstein in Spain*...op. cit., pp. 8-11.

[55] SANCHEZ-RON, J.M. (ed.) *La Junta para Ampliación de Estudios e Investigaciones Científicas 80 años después, 1907-1987.* CSIC. Madrid 1988. (2 vols.). SANCHEZ-RON, J.M. (1988) "La física en España durante el primer tercio del siglo XX" a SANCHEZ-RON, J.M. (ed.) *Ciencia y sociedad en España.* op.cit. MILLAS VALLICROSA, J.M. "La vindicación de la ciencia española por Menéndez y Pelayo," *Arbor*, 127-128, julio-agosto 1956, 410-426. SALA CATALA, J. "Ciencia biológica y polémica de la ciencia en la España de la Restauración" en SANCHEZ-RON, J.M. (ed.) (1988) *Ciencia y sociedad en España.* op. cit. pp. 157-177.

[56] LAPORTA, F.J., RIUZ MIGUEL A., ZAPATERO, V.,SOLANA, J. "Los orígenes culturales de la Junta para Ampliación de Estudios," *Arbor*, 493, 1987, 17-87. In particular "Una interpretación del problema de la decadencia española" pp. 35-47.

[57] LAPORTA, F.J., RIUZ MIGUEL A....op. cit. p. 36.

[58] COMAS SOLA, J. "Nuestra decadencia" *La Vanguardia*, 28 November 1899. Cited in GARCIA CAMARERO, E. y E. (eds.). *La Polémica de la ciencia española.* op. cit. pp. 400-409.

[59] "...en España, además de la poca instrucción que existe (¡se cuentan doce millones de españoles, en la Península, que no saben leer!...), entre los pocos científicos que hay, casi es nulo el espíritu de investigación y el amor al trabajo y al estudio...lo que falta es ciencia teórica y práctica, amor al progreso, cumplimiento de su deber y espíritu de mejoramiento". COMAS SOLA, J. "Nuestra decadencia" op. cit. pp. 403-404.

[60] PESET, J.L. "Pedro Laín y la polémica de la ciencia española," *Arbor*, 562-563, 1992, 27-34.

[61] PESET, J.L. "Pedro Laín y la polémica de la ciencia." op. cit., p. 29.

[62] LORA TAMAYO, M. "El momento actual de la ciencia española" *Arbor*, 1949, 43-44, 381-393.

[63] ALBAREDA, J.M. "La aptitud investigadora y otros factores de la producción científica" *Arbor*, 60, 1950, 337-355.

[64] ALBAREDA, J.M. "La aptitud investigadora.." op. cit. p. 340.

[65] "Lamento decir que el ambiente científico español no ha cambiado. No es un problema de orgnización sino de falta de ideas. No observo ningún movimiento que nos permita pensar que estamos en un momento de renacimiento de la ciencia española," *El País*, 29-V-1985. Cited by MARQUEZ, A. (ed.) *Ciencia e Inquisición. Arbor*, 1986, 24 (Introduction, p. 24).

[66] This could be the case of: LOPEZ PIÑERO, J.M. *Ciencia y Técnica...*op. cit. CAPEL, H., SANCHEZ, J.E., MONCADA, O. *De Palas a Minerva. La formación científica y la actividad espacial de los ingenieros militares en el siglo XVIII.* El Serbal/CSIC. Barcelona 1988. CAPEL, H. *Geografía y Matemáticas en la España del siglo XVIII.* Oikos-Tau. Barcelona 1981. This idea is also clearly expressed in Antonio Lafuente's quotation in an article on the Spanish scientific institutions in the eighteenth century: "The principal object of this study is to try to describe the unique organizational structure of Spanish science in the Enlightenment...it seems, on the one hand, not to have contributed anything to scientific development, but at the opposite extreme Spanish science made notable contributions to eighteenth-century culture in areas such as geography, botany and anthropology" LAFUENTE, A. "The metropolitan institutionalization of Spanish Scientific Activity", unpublished manuscript.

[67] LOPEZ OCON, L. "Ciencia e historia de la Ciencia.." op. cit. pp. 102-103.

[68] RODRIGUEZ CARRACIDO, J. *Estudios histórico-críticos...*op. cit.

[69] RODRIGUEZ CARRACIDO, J. *La Evolución en la química.* Vola. De Hernando. Madrid 1894.

[70] Alexander von Humboldt, the botanist Francesc Xavier Bolòs (1773-1844), the chemists Proust, García Fernández, the mineralogist Christian Herrgen, or the life of Jovellanos, a minister of Charles III. RODRIGUEZ CARRACIDO, J. *Jovellanos. Ensayo dramático-histórico*. Fortanet. Madrid 1893.

[71] MASCAREÑAS, E. op. cit.

[72] RUSSELL, C.A. "Rude and Disgraceful Beginnings': A view of History of Chemistry from the Nineteenth Century" *British Journal for the History of Science*, 21 (1988), 273-294.

[73] MOLES, E. *Del momento científico español 1775-1825. Discurso leido en el acto de su recepción en la Academia de Ciencias Exactas, Físicas y Naturales de Madrid por E. Moles...el dia 28 de marzo de 1934*. Madrid 1934. NIETO-GALAN, A. "Enric Moles i Ormella (1883-1953): La importació d'una nova disciplina, la química-física" in A. ROCA-ROSELL, A., CAMARASSA, J.M. (eds.) *Ciència i Tècnica a l'època contemporània als Països Catalans: Una aproximació biogràfica*. FCR. Barcelona, 1995. II, pp. 1147-1176.

[74] NIETO-GALAN, A., "Seeking an identity for chemistry in Spain: Medicine, Industry, University, the Liberal State and the new 'Professionals'" in KNIGHT, D., KRAGH, H. (eds.) *The making of the chemists in nineteenth-century Europe*. CUP. Cambridge 1998 (in press).

[75] LOPEZ PIÑERO, J.M. *Ciencia y Técnica*...op. cit. p. 20.

[76] HERNANDEZ MOREJON, A. *Historia bibliográfica de la medicina española*. Vda. de Jordán e Hijos. Madrid 1842-1852.

[77] MENENDEZ PELAYO, M. *La Ciencia Española*. CSIC. Santander 1953 (5th edition, especially issued to reconstruct the history of the "polemica"). (3 vols.)

[78] PEROJO, J. "La ciencia española bajo la Inquisición", op. cit.

[79] LOPEZ PIñERO, J. M. et. al. *Diccionario histórico de la Ciencia moderna en España*. Península. Barcelona 1983. (2 vols.). LOPEZ PIñERO, J. M. et. al. *Bibliographia Medica Hispanica*. Instituto de Estudios Documentales e Históricos sobre la Ciencia. Valencia 1987-1990. (8 vols.).

[80] LOPEZ PIÑERO, J.M. *Ciencia y Técnica*...op. cit. (Introduction).

[81] LOPEZ PIÑERO, J.M. *Ciencia y Técnica*...op. cit.

[82] MILLAS VALLICROSA, J.M. *Estudios sobre historia de la ciencia española*. CSIC. Barcelona 1949. MILLAS VALLICROSA, J.M. *Nuevos estudios sobre historia de la ciencia española*. CSIC. Barcelona 1960.

[83] MILLAS VALLICROSA, J.M. "La vindicación de la ciencia..". op. cit. pp. 411-412.

[84] LOPEZ PIÑERO, J.M. *Ciencia y Técnica en la sociedad española de los siglos XVI y XVII*. Op. cit. ESTEBAN PIÑEIRO, M., "Cosmografía y matemáticas en la España de 1530 a 1630" *Hispania*, 177, 1991, 329-337. GOODMAN, D. *Power and Penury. Governement, Technology and Society in Phillip II Spain*. CUP 1988. (Spanish translation 1990). GOODMAN, D. "The Scientific Revolution in Spain and Portugal" in PORTER, R., TEICH, M. (ed.) *The Scientific Revolution in National Context*. CUP. Cambridge 1992, 158-177.

[85] MILLAS VALLICROSA, J.M. "La vindicación de la ciencia.. op. cit. p. 422.

[86] VERNET, J. *Historia de la Ciencia española*. op. cit. SAMSO, J. *Islamic astronomy and medieval Spain*. op. cit.

[87] LAFUENTE, A. SELLES, M., PESET, J. L. op. cit.

[88] MARQUEZ, A. (ed.) *Ciencia e Inquisición*. Arbor, 1986, 24 (Introduction, p. 24)

[89] LOPEZ PIÑERO, J.M. "Las etapas iniciales de la historigrafia de la ciencia. Invitación a recuperar si internacionalidad y su integración" *Arbor*, 142, 1992, 21-67. LOPEZ PIÑERO, J.M. "La tradición de la historiografia de la ciencia y su coyuntura actual. Los condicionantes de un congreso" in LAFUENTE, A, ELENA, A, ORTEGA, M.L. (eds.) *Mundialización de la ciencia y cultura nacional.* Actas del Congreso Internacional "Ciencia, descubrimiento y mundo colonial". Doce Calles. Madrid 1993. pp. 23-49.

[90] A useful source of approaches of this kind of approaches can be found, for example, in the volumes published after the regular meetings of the Spanish Society for the History of Science and Tecnology.

[91] ROCA-ROSELL, A. "Una perspectiva de la historiografia de la ciència i de la tècnica a Catalunya" in NAVARRO, V. et. al. (eds.) *Actes de les II Trobades d'Història de la Ciència i de la Tècncia.* IEC. Barcelona 1993. pp. 13-26. NADAL, Jordi. *El fracaso de la revolución industrial en España. 1814-1913.* Ariel. Barcelona 1988. From the point of view of the history of technology in the Rennaissance and the links with the Spanish colonies, see: GARCIA-TAPIA, N. *Técnica y Poder en Castilla durante los siglos XVI y XVII.* Salamanca 1989. GARCIA-TAPIA, N. *Del Dios del Fuego a la Máquina de Vapor. La introducción de la técnica industrial en Hispanoamérica.* Ambito. Valladolid 1992.

[92] ROCA-ROSELL, A-, SANCHEZ-RON, J.M. *Esteban Terradas (1883-1950). Ciencia y Técnica en la España Contemporánea.* INTA. Barcelona 1990.

[93] " [La ciencia española] depende de la aventura de hombres aislados, o de grupos determinados capaces de conectar con Europa y de contribuir a importantes proyectos internacionales pero trabajando siempre contra la indiferencia de su propia sociedad," LOPEZ PIÑERO, J.M., "La marginación de la ciencia en la España contemporánea" in GONZALEZ BLASCO, P., JIMENEZ BLANCO, J., LOPEZ PIÑERO, J.M. *Historia y sociologia de la ciencia en España.* Alionta. Madrid 1979, pp. 72-93. p. 77.

[94] COOTER, R., PUMFREY, S. "Separate spheres and public places: reflection on the history of science popularozation and science in popular culture," *History of Science*, 1994, 32, 237-267. PUMFREY, S., ROSSI, P., SLAWINSKI, M. (eds.) *Science, Culture and popular belief in Renaissance Europe.* Manchester University Press. Manchester 1991. RHYS MORUS, I. "Manufacturing nature: science, technology and Victorian consumer culture," *British Journal for the History of Science*, 1996, 29, 403-34. SHAPIN, S. "Science and the public" in OLBY, R.C., et al. (eds.) *Companion to the History of Modern Science.* op. cit. pp. 990-1007. COOTER, R. *The cultural meaning of popular science:phrenology and the organization of consent in nineteenth century Britain.* CUP. Cambridge 1984. TURNER, F.K. "Public Science in Britain, 1880-1919," *Isis*, 71 (1980), 589-608.

[95] SCHROEDER-GUDEHUS, B. "Nationalism and internationalism" op. cit.

[96] A good example is: GLICK, Th. *Einstein y los españoles. Ciencia y sociedad en la España de entreguerras.* Alianza. Madrid 1986.

[97] STEWART, L. *The Rise of Public Science: Rhetoric, technology and natural philosophy in Newtonian Britain, 1660-1750.* (Cited in COOTER, R., PUMFREY, S. "Separate spheres and public places:.." op. cit. p. 253.)

[98] In the case of the United States see for example: SMITH, M.R. "Technological Determinism in American Society" in SMITH, M.R., MARX, L. *Does Technology drive History?. The Dilemma of Technological Determinism.* MIT Press. Cambridge,MA. 1994. pp. 1-36.

[99] BLOOR, D. *Knowledge and social imagery.* London, 1976, BARNES, B. SHAPIN, S. (eds.), *Natural order: historical studies of scientific culture.* London, 1979. BIJKER, W.E., HUGHES, T.P., PINCH, T.J. (eds.), *The Social Construction of Technological Systems. New Directions in the Sociology and History of Technology.* Cambridge, Mass. 1987. For a

general historiographical review see: FOX, R. (ed.) *Technological Change. Methods and Themes in the History of Technology*. Amsterdam 1996.

[100] For the history of Technology, see: W.E. BIJKER, T.P. HUGHES, AND T.J. PINCH (eds.), *The Social Construction of Technological Systems. New Directions in the Sociology and History of Technology*. MIT Press, Cambridge, Mass 1987.

[101] MORREL, J., THACKRAY, A. *Gentlemen of Science. Early Years of the British Association for the Advancement of Science*. OUP. Oxford, 1981, p. 12.. For a general introduction to the culture of Victorian Britain see: ALTICK, R.D. *Victorian People and Ideas*. J.M. Dent. London 1974. CANNON, S. *Science in Culture: The Early Victorian Period*. Dawson. New York 1978.

[102] I would like to thank Dr. Antoni Roca-Rossell (Universitat Politècnica de Catalunya), Dr. Xavier Roqué (Universitat Autònoma de Barcelona), Dr. Jon Arrizabalaga and Dr. José Pardo Tomás (CSIC, Barcelona), and Prof. Kostas Gavrouglu (University of Athens) for their generous comments on the first draft of this paper. Arquimedes' language editing has also been of great value. The translation of the Spanish quotations into English are mine.

LUIGI CERRUTI

DANTE'S BONES

Geography and History of Italian Science, 1748-1870

> Nous serions injustes si [...] nous ne reconnaissions point ce que nous devons à l'Italie; c'est d'elle que nous avons reçu les sciences, qui depuis ont fructifié si abondamment dans toute l'Europe; c'est à elle surtout, que nous devons les beaux-arts et le bon goût, dont elle nous a fourni un grande nombre de modéles inimitable.
>
> *Jean Le Rond d'Alembert, Discours préliminaire, 1751*

This essay will outline a profile of Italian science, against the background of European science, in the period between the Peace of Aachen of 1748 and the annexation of Rome to a unified Italy in 1870. The significance of the choice of two political events to demarcate the study will become clear in the course of the essay, but the dates have a precise historical meaning. After the Peace of Aachen the Peninsula remained fragmented into many States, which nevertheless had stable borders, and for half a century it enjoyed an unusual peace. In 1870 the lay bourgeoisie - and the aristocracy - conquered that Italian city that had for two millennia been a 'universal' symbol of power. The two events mark the beginning and end of a notably long period, a historiographical choice that permits the individuation of long-standing relations within and outside the Italian scientific community. At the same time, the geographic heterogeneity of the Peninsula means that in order to follow the vicissitudes of Italian science we shall have to adopt a somewhat ad hoc - although quite specific - interpretation of the center/periphery opposition. It is not at all easy to give a strict definition of what constituted 'Italy' in the period considered. Further difficulties are added if we ask who, at that time, could be defined as an 'Italian scientist'. The two questions will be treated separately.

A LITERARY DEFINITION OF ITALY, OR LOOKING FOR A CENTRE

Between 1772 and 1781 a major work on the Italian cultural heritage was printed in Modena. The author was Girolamo Tiraboschi (1731-1794), a Jesuit since 1746, who had used the huge collections of the Biblioteca Estense, in Modena, to write a "history of the origin and of the progress of all the sciences in Italy", i.e. "in that stretch of country (*tratto di paese*) that now is called Italy".[1] The erudite Jesuit treated poetry, fine arts, medicine, law, sacred eloquence, philosophy and mathematics from the most remote antiquity until (his) present.[2] Several aspects of this *Storia della letteratura italiana* are very pertinent to our historiographical interests. The *container* of Tiraboschi's material is, in a sense, Italy itself, as the "geographical definition" of the "solid container of the fluid matter that he was interested to collect".[3] On general grounds, Tiraboschi proposed a polemic contrast between the massive and immemorial Italian tradition and the traditions of other peoples, a sort of veiled echo to d'Alambert's posthumous recognition in his *Discours sur l'Encyclopédie*. In the opening of the second tome of the revised edition of his *Storia*, Tiraboschi discussed four topics, in this order: the problem of decadence, the nature of the Italian language, the Copernican system, and the conviction of Galileo. The last two topics were of course of a certain weight for a Jesuit. The theme of decadence,[4] it should be noted, was one of the great problems of the XVIII century, and a typically European one, with the Frenchmen J.B. Du Bos and Montesquieu as principal references. In this context, Tiraboschi distinguished between a possible decadence of humanities and fine arts and the existence of irreversible advances as, for example, the discovery of the laws of the vacuum.[5] Tiraboschi's was clearly a 'progressive' response to the problem. On the question of the aptitude of the contemporary Italian tongue to express all the nuances of the modern culture, literary but also scientific and technical, Tiraboschi attempted to refute the criticism of his Spanish brother Jesuit, Esteban Arteaga, but was in difficulty because of the lack of Italian works in many sectors of the current literature.[6]

Following up the cultural interests of Tiraboschi, a crucial knot appears: the historical role of the Italian language in the very definition of the 'nation' called Italy. In the XVIII century the Italian culture and language was 'invaded' by French civilisation, and many Italian intellectuals gave their contribution to this intrusion simply as active advocates of the *lumi*, the ideas of the Enlightment. One of the most important groups of Italian illuminists was active in Milan and published a journal with the deliberately bourgeois title *Caffè*. In the issue of July 1764, Count Alessandro Verri (1741-1816) published an article, in which he and his friends declared that they had formally, "in the presence of a notary", renounced the "alleged purity of the Tuscan idiom (*toscana favella*)".[7] In fact, the question of the language *in use* in the Country was very serious, from many points of view, cultural as well as political. In 1785 Melchiorre Cesarotti (1730-1808), professor of Greek and Hebrew literature in Padua, plainly stated that "the use gives the law", and that in "a nation divided in different provinces, without a capital that has any monarchic jurisdiction over the other capitals", no dialect (such as that of Florence) should be the norm for the common language.[8] In 1791 Count Gianfrancesco Galeani Napione (1748-1830), from the border region of Piedmont, wrote against Cesarotti, publishing a treatise which defended the use of Italian instead of Latin or French, and which was, at the same time, a vindication of the Italian tongue as a symbol of national pride.⁹ But since 1766 the point had been politically very explicit in the words of Saverio Bettinelli (1718-1808), once more a Jesuit. "In Italy every province has a Parnassus, a style, a taste, and, along with the genius of the climate, a party, a league, a judgement separated from the others", "it was pleasant to change nation and habits at every change of horses [...] but I was irritated by never knowing where Italy was". The conclusion of this train of thought was important: "if Italy really had a centre, a point of union, she whould be richer in the fine arts, in the letters, and perhaps in the sciences; much more than any other nation".¹⁰ This *search for a centre* was proposed by Bettinelli in his *Lettere inglesi*, written in the context of a brisk polemic about Dante, which he himself had launched in 1757 with his *Lettere virgiliane*. This was a very vigorous pamphlet, so vigorous that it shocked many liberal minds, such as that of Gaspare Gozzi (1713-1786), a clever Venetian *letterato*, who wrote a *Difesa di Dante* (1758). As a result of this and further polemic about Dante, the Florentine poet definitively became the national Italian poet, thus reinforcing the claim of the Tuscan language of Dante and Boccaccio to be adopted as the written language of all of Italy;¹¹ nevertheless, even the final resting place of Dante's bones remained questioned for many years after the Unity of Italy. A look at the strange fate of Dante's remains may give us some more hints on the Italian mind (and a first explanation of the paper's title).

Dante Alighieri died in Ravenna during the night of 14 September 1321, and in the same town he was buried with great ceremony in the Chapel of the Madonna, near a church known at that time as San Pier Maggiore. In 1483 an elegant monument by Pietro Lombardo (1435-1515) replaced the original, simple marble sarcophagus. The church belonged to the Franciscan Friars, so that they began to consider the remains of the great poet as the property of the Order. At some point during the following centuries, however, the Friars felt their precious asset to be threatened by other powerful claimants, and for safekeeping the bones were removed from the tomb by a Friar named Santi, collected in a small rudimental urn and immured in a wall close by.¹² The urn and bones were found by chance on 27 May 1865, when the Chapel was being refurbished in the occasion of the sixth centenary of Dante's birth. This small, unhappy story becomes paradigmatic in the light of the two alternative explanations that historians have proposed for the concealment. In the lay, political version, the cause was the ceaseless pressure on Ravenna by Florence, to bring Dante's remains back to his birth place; it was during negotiations between Lorenzo de' Medici and Bernardo Bembo in 1475-76 that the Friar decided to move the bones to a safer place.¹³ The other version holds that the entire affair developed within the Roman Catholic Church, with the Franciscan Order on one side and the Papal Delegate Domenico Corsi on the other, struggling in 1677 for the possession of the bones. The pressure from Rome became so strong that the Order felt itself compelled to hide them.¹⁴ It is to be added that Ravenna is about

160 kilometres north-east of Florence; this moderate physical distance represented a political and cultural gap that could not be overcome in the XV century, nor in the XIX century, when, after the finding of the urn, Florence again asked Ravenna to return Dante's bones. No highest national interest was strong enough to induce Ravenna's people to give up their treasure to Florence.

The widespread admiration for Dante, growing after the second half of the XVII century, did not improve the practical fortunes of the Italian language. The historians of Italian literature and language are unanimous in describing two parallel cultural processes: on one hand the growing belief that the Country's great cultural past gave its intellectuals the task of fighting for the independence, unity, and liberty of Italy; on the other hand the formation and spread of a national language was very slow, because a more or less 'regional' Italian was really spoken only by an *élite* - and often not in every-day life even by this *élite*. To be sure, that sort of literary nationalism did not wait for the French Revolution to be strongly affirmed in books, pamphlets and, especially, in private letters. This, for example was the case of Saverio Bettinelli, Giuseppe Baretti (1719-1789) and Alessandro Verri, who exalted the love of freedom as a characteristic trait of Italians.[15] A Piedmontese writer, Vittorio Alfieri (1749-1803), worked for many years in this direction on a sort of manifesto, published exactly in 1789. He explicitly connected his own thought to that of Machiavelli, and in many pages of his *Del principe e delle lettere* he urged the liberation of Italy from the foreign "barbarians", and predicted the unity of the Nation.[16] The historian of Italian literature Walter Binni has defined this work as "fiery pre-romantic pages that mark the violent birth of the Italian national sentiment".[17] About twenty years after Alfieri's pamphlet, in 1806, Alessandro Manzoni (1785-1873), wrote to a friend, in a quieter and concerned mood: "To our misfortune, the state of Italy, divided into fragments, the laziness and the almost general ignorance, have put so great a distance between the written language and the spoken one, that this may almost be called a dead tongue". Manzoni's truly great contribution was the transformation of the question of the language from a quarrel between men of letters into a civil problem for the Italian nation.[18] A national problem indeed, with many different facets, like the use of dialects even by the ruling classes in the various Italian regions. Ugo Foscolo (1778-1827), the great poet and patriot, bitterly affirmed that, while in the other European countries "educated people" used the "national tongue" and left the dialects to the populace, in Italy the use of Italian was permitted only to travellers in nearby provinces: "a language which is common only so far as it is necessary to be understood, and that could be named *mercantile* and *itinerant* (*itinerario*)".[19]

During the XVIII century the linguistic situation of the Italian *élite* has been described by a contemporary Italian linguist as a "new bilingualism", with French as the language of culture.[20] The consequences were particularly serious for the publishing trade, a crucial point for the development of culture. For example, in Italy *Iphigénie* had 15 editions between 1708 and 1799,[21] and the huge *Encyclopédie* was reprinted twice *in French* (albeit with notes to soften the anti-Christian trend of the French enlightenment).[22] At the same time, in publishing centres such as Venice it was lamented that works written in Italian had no market, and so they were not printed at all.[23] It is obvious that this distressing situation was of great importance not only for men of letters, but also for the other *letterati* whom we now call scientists. I will return several times to this and to other linguistic aspects of the relationship between the Italian scientific centres and the more or less prestigious centres abroad.

Before leaving this first look at the literary definition of 'Italy' - the only definition feasible at that time - it may be useful, since Piedmont was the nurturing place of Italian Unity, to look more closely at the situation in that region before the Unification itself. It is to be remembered that Piedmont became the 'centralized model' for the rest of Italy the after its annexation to the hereditary kingdom of Savoy. The Turinese Giuseppe Baretti was one of the most cosmopolitan intellectuals of the XVII century, but about his linguistic commitment Devoto wrote that he was "in general dominated by the habit, characteristic in a Piedmontese, obsessed by problems of language, of aiming at a distinguished (*illustre*) vulgar tongue".[24] A generation later, Alfieri tried to speak the Florentine dialect fluently and, when in Florence he had Piedmontese guests, he alternated Florentine (with a careful pronunciation) with Piedmontese. In 1786 he asked a friend to send him a secretary, waiter, and servant from Siena, in order to have all around himself "pieces of living vocabulary".[25] However, not all the Piedmontese intellectuals were so obsessed. In 1803 another important Piedmontese intellectual, the liberal clergyman and historian Carlo Denina (1731-1813), proposed - from Berlin - the use of French as the general language of culture in the Piedmont which was at that time annexed to France.[26] Probably Denina was looking clear-sightedly at the reality of his birth place. Alfieri himself, and Galleani Napione, recognized (or deplored) that in Turin the upper classes almost exclusively used dialect or French.[27] The situation did not improve at all after the Restoration: in 1831, strolling through the streets of Turin or sipping a *café*, a traveller heard people speaking only dialect or French. The bookseller and publisher Pomba declared that, only French was written and read in the city, and that books in Italian remained unsold. Even the soldiers were commanded in French.[28] Most impressive of all the witnesses, Marquise Arconati said that, after 1848, when the members of the Turin Parliament tried to speak Italian it was clear that they were speaking a dead tongue, which they were not in the habit of using.[29]

In conclusion, it is important to note the deep social and historical contradiction between the widespread awareness of the overwhelming unifying significance of the cultural heritage expressed by the *literature in Italian*,[30] and the very visible fact that the same language, in which this treasure was stored up, was not the common, 'usual' property of Italians. At the end of the process of unification, it has been calculated that out of 25 million Italians, only 750,000 were users of the literary tongue.[31] The goal of transforming "a *cultural* nation into a *territorial* nation"[32] had been reached, but another objective, equally important, had been badly missed, or rather had never been sought. I am referring, with Migliorini, to the fact that in the first half of the XIX century the number of Italians wishing for *territorial unity* quickly grew to reach the impetus of an avalanche, but the *social unity* of the Nation remained (and in my opinion still remains) the aim of only few.[33]

SCIENCE IN ITALY, ITALIAN SCIENCE, AND ITALIAN SCIENTISTS

Science in Italy and Italian science: it is immediately clear that these expressions are not at all equivalent. After the preceding discussion about the literary definition of Italy, the expression *science in Italy* may appear very approximate; in the period considered here it is actually wrong. One of the most important scientific centres of that period was the University of Pavia, and, apart from the relatively short period of Napoleonic domination, it was until 1859 a University of the Austrian Empire. As we will see, the consequences of this situation for the science 'produced' in Pavia depended strongly on what happened in Vienna, and on the political relationships between the Austrian capital and the Italian provinces. More sadly, for several reasons the question of whether the States of the Church were really in the Peninsula is by no means rhetorical, because serious doubts would be raised by a positive answer, not in regard to space, but in regard to time. To put it in a nutshell, medieval astronomy officially ended in Rome only between 1822, when the condemnation of the Copernican theory was cancelled, and 1835, when the works of the Polish astronomer were removed from the new edition of the *Index librorum prohibitorum*.[34] From the historiographical point of view, the expression *Italian science* is somewhat better; I use it in the title of this paper because the adjective 'Italian' means "relating to Italy *or* its people or language."[35] Italian people and Italian language existed many centuries before territorial Unity and the proclamation of the Kingdom of Italy in 1861; at the same time, an Italian culture existed, definable anthropologically as the customs, civilisation and achievements of the Italian people. Thus I will speak of Italian science as an integral part of Italian culture; however, the term 'Italian' must always deeply scrutinized when it is used in a political, or, as here, in a historiographical context.

The 'anthropological' definition of culture just mentioned is in fact gravely defective, because on the basis of the presently available historical documentation, all the most important terms of the cultural discourse - customs, civilisation, achievements - refer only to a thin layer of the people living in the Peninsula. The Italian people were 'discovered' at the time when the French revolutionary army was invading the Northern regions of the Peninsula, and the most discerning moderate intellectuals, such as the Lombard Count Paolo Greppi, saw clearly the need for mass enrolment of the peasants, to be paid for by moderate reforms. Before the time of the Revolution the Italians had been "frozen into masked figures, classified into vernacular varieties, confused with the vegetation and the ruins of the landscape". Unfortunately these terrible words[36] correspond exactly to the view shared by almost all the intellectuals of the period, and in large part also to the current state of our knowledge. From the Restoration onward, the Italian people were always urdergoing 'refurbishing': during the *Risorgimento*, after Unity, during the First World War, under the Fascist Regime, right up to the 'work in progress' at the time of writing this essay. A generation ago, Giulio Bollati affirmed that "the trend toward the industrial production of popular feelings" started because of "the need for consensus, which was absolutely necessary to the bourgeois revolutions and to the transformation of the old dynastic States into a mass nation".[37] At a few points, my research will marginally touch these questions also, but it was again necessary to stress the ambiguity of the term *Italian*, which, nevertheless, I will need to use in every passage of the paper.

Different and milder problems are posed by the expression *Italian scientist*, which, nevertheless, is also far from innocent in the historiography of science. Before giving a tentative interpretation, it will be useful to go through a few examples, to varying degrees controversial. The following discourse about the *Italian-ness* of scientists will concern Boscovich, Lagrange, Avogadro, Malaguti, Moleschott, Dohrn. They will provide a first look at some crucial aspects of Italian science in the period considered.

Ruggiero Giuseppe Boscovich (1711-1787), a Jesuit, had an important role in the science of his time, but his influence was perhaps more intense long after his death, in the physics and chemistry of the XIX century. Everyone knows that Boscovich was born in Ragusa on the Dalmatian coast, but to know what kind of place Ragusa was then, we may look at it through the eyes of a *letterato* of the time. Alberto Fortis (1741-1803), a naturalist from Padua, had began the 'exploration' of the East coast of the Adriatic with a journey by ship to two islands, Cherso and Osero, a few hours sailing from the *Dominante* (an appellative of Venice, of course). He was "in a small party with Mr. Giovanni Symonds, an English gentleman and with Dr. Domenico Cirillo, professor of botany and natural history in Naples",[38] and they were fascinated by the diversity of natural and human habitat found there. Before following Fortis on this journey towards Ragusa, it is interesting to note that in the second part of the XVIII century such an important part of Italy as Venice required journeys of exploration in order to know itself. When Fortis arrived in Ragusa he wrote: "The spirit of literature that flourishes in Ragusa surprises every thoughtful man. The inhabitants of this small capital do not amount to nine thousand".[39] The small town had a notable merchant fleet, one of the most important in the Mediterranean,[40] and had been independent of Venice for several centuries. It acknowledged the Ottoman Empire sovereignty, but only with the payment of a levy. Germano Paoli's thorough historical analysis demonstrates that Ragusa was an Italian Merchant Republic, as is demonstrated, for instance, by its official international correspondence in Italian.[41] Boscovich corresponded with his relatives in Italian, lived in Italy for 52 years of his life, and was an Italian member of the scientific society founded in 1781 by Lorgna (see below); however, when he died on 13 February 1797 in Milan, many connected his genial and many-sided activity with the autonomous republican activities of his nation, Ragusa. On many occasions he demonstrated a "perspicacious will for conservation", that also may be related to the culture of the Merchant Republic.[42] In Boscovich's correspondence, Venturi found many references to the *amor di patria*.[43] In the course of a harsh 'scientific' attack, d'Alembert had referred to Boscovich as *un géometre italien*, in his 'scientific' answer the Jesuit states: *notre Auteur* (himself) *est Dalmate et de Raguse, non Italien,* but at once adds: *Cependant vu le long séjour qu'il a fait en Italie depuis sa premiére jeunesse, on peut en quelque sorte le dire Italien.*[44] It should also to be mentioned that Boscovich was compelled to share the fate of his order when Pope Clement XIV decreed the suppression of the Order of Jesuits in July 1773. He repaired to France, to Paris, where he was appointed to an especially-established post, that of *Directeur d'Optique au service de la Marine*.[45] He was received with mixed feelings by the 'local' *uomini di lettere*, and he had to pay 1500 francs for the citizenship necessary for a civil servant.[46] He was able to return in Italy in October 1785, and died in Milan. The argument about Boscovich's nationality is still flourishing, although his biography is in the *Dizionario Biografico degli Italiani*.[47]

Giuseppe Ludovico De la Grange Tournier (1736-1813) was born in Turin, the capital of the Kingdom of Sardinia. Whereas the young Boscovich, on the far Eastern border of the Italian cultural zone, could look beyond the border to the Ottoman Empire, Lagrange in Turin looked Westwards, to France. Lagrange had a vivid, profound mathematical intelligence; he was nineteen years old when, in a letter to Euler, he communicated the principle of the variation calculus, and in the same year, 1755, he was appointed teacher of mathematics at the Turin School of Artillery. In 1759, he and two other young *letterati*, Count Giuseppe Angelo Saluzzo di Monesiglio (1734-1810) and Giovanni Francesco Cigna (1734-1790), founded the Società Privata Torinese, the original nucleus of the Turin Academy of Science. His fame as a mathematician grew quickly throughout Europe, and he received offers from many countries; the most attractive came from Berlin, when Euler moved from there to Petersburg, and on 21 August 1769 Lagrange left Turin for the scientific court of Friedrich II. The voluntary 'exile' of a *letterato* of fame from Piedmont was not exceptional, and as an example I may recall Baretti and Denina, whom we encountered in the preceding section, discussing the fate of the Italian language, writing from London and Berlin respectively.

In Berlin, Lagrange did not feel German. Avoiding local entanglements, he summoned to Berlin and married a cousin, Vittoria Conti; in a similar spirit, when Lorgna proposed him as a member of the Società Italiana delle Scienze, he answered that he was "anxious to merit this honour and, at the same time, to show himself a good compatriot (*buon compatriota*)".[48] We are also sure that Lagrange was not French, because when he was elected a member of the Parisian Académie des Sciences he became one of the eight *foreign* members of the Academy.[49] On the death of Friedrich II the Italian mathematician received offers from the Kingdom of the Two Sicilies, the Kingdom of Sardinia, and the Grand-Duchy of Tuscany, but in 1787 he preferred to go to Paris. In the French capital his nationality did not change abruptly: when all aliens were banished from France during the Terror, on 16 October 1793, Lagrange was only able to remain in Paris thanks to a special decree.[50] In the following years, the question of the citizenship of the great mathematician lost all meaning, because since 1798 all the inhabitants of Turin, occupied by the French troops, also became *citoyens*. In 1802 Lagrange himself proposed the project for constitutional revision that formally annexed Piedmont to the French Republic; the scientific work of the Piedmont-born mathematician could not be annexed so easily. In 1810, in a *Discorso sui recenti progressi dovuti agl'italiani delle scienze matematiche e fisiche* [Discourse on the recent advancements due to Italians in the mathematical and physical sciences] Lagrange was claimed as Italian by the intellectuals of the Istituto Nazionale del Regno d'Italia, both of which, the Institute and the Kingdom, had been established by Napoleon.[51] But, of course, none could doubt that Lagrange was a French mathematician when he signed the 1811 edition of his great *Mècanique analytique*, with a fully fledged *par J.L. Lagrange, de l'Institut des Sciences, Lettres et Arts, du Bureau des Longitudes; Membre du Sénat Conservateur, Grand-Officier de la Légion d'Honneur, et Comte de l'Empire.*[52]

Lagrange died on 10 April 1813; Napoleon was defeated in the so-called Battle of the Nations at Leipzig on 16-19 October of the same year. The silences and the laments on Lagrange's death were marked by the changing borders of Napoleon's power. The Universities of the Regno d'Italia (Pavia, Bologna, Padua) could mourn the great *géomètre*, but Lagrange - an enemy - was not commemorated by the Berlin Academy; the commemoration by Anton Maria Vassalli Eandi (1761-1825) at the Turin Academy was read on 3 May 1813, but after the Restoration of Savoy (20 May 1814) it was never printed. The most celebrated contemporary biography is that by J.B.Delambre, *Secrétaire perpetuel* for the mathematical sciences at the Institut de France. Here Lagrange was completely 'annexed' to France, owing to his surname, his mother, the language used in his works.[53] We need not to go through the long quarrels about Lagrange's nationality, but an almost 'spontaneous' episode is to be remembered, because the judgement was expressed by a man relevant to the history of Italy, within whom there was an interesting interplay between physical-mathematical style and 'nationalistic' motivation. Count Luigi Federico Menabrea (1809-1896) was born in Chambéry,[54] the old capital of Savoy, and cradle of the dynasty of the same name. He was a good physicist, interested in the theory of elasticity, an officer of the Piedmontese Army, and would finally be Prime Minister of unified Italy in the years 1867-1869, when Chambéry had for several years been a French town. In 1846, the Eighth Meeting of Italian Scientists[55] was held in Genoa, and there Menabrea had a lively discussion with a pupil of Avogadro's, Felice Chiò (1813-1871). The discussion was about the validity of a certain demonstration against some points of Lagrange's theory of series. Chio's results had been refused by the Academy of Turin, which was a stronghold of the Lagrangian tradition, but, thanks to Cauchy, Chiò had begun to publish his papers in Paris.[56] In the course of the discussion, Menabrea defined Lagrange as "the immortal Turinese geometrician" and "the immortal Italian geometrician", looking for his own scientific roots as well as for the 'construction' of an all-Italian scientific tradition.[57]

As far as I know, Avogadro's 'nationality' is not now a question, but Amedeo Avogadro (1776-1856), like Boscovich and Lagrange, was never an Italian citizen, because he died before the proclamation of the Kingdom of Italy in 1861. To be sure, during his life an earlier Kingdom of Italy had been born, when Napoleon crowned himself King of Italy in Milan (May 1805), but by then Avogadro had for several years been a French citizen, and was employed as a French civil servant, the *secrétaire du départment* in the Turin prefecture.[58] In 1811, when his *Essai d'une manière de déterminer les masses relatives des molécules élémentaires des corps* was published, he was a teacher of 'positive philosophy' (mathematics and physics) at the Royal College of Vercelli, a small Piedmontese town; we may note that his famous essay was written in French by a French citizen, and published in a French journal.[59] All of Avogadro's papers published between 1806 and 1814 were printed in French journals; only after the Restoration did the Turinese physicist begin to publish in Italian.

Up to here the whole story may seem a *mot d'esprit*, but I must add that only by the efforts of Ampère's 'invisible school' was Avogadro's hypothesis able to survive in the scientific community of the first half of the XIX century, and was finally able to strike the imagination of Cannizzaro.[60] Incidentally, Cannizzaro preferred to call the fundamental law on gases Avogadro's *and* Ampère's hypothesis.[61]

The life of Faustino Malaguti (1802-1878) followed yet another scenario. He was born in the States of the Church, near Bologna, and in the University of that town he took his pharmacist's certificate at the early age of sixteen. There are no traces of scientific activity before the event that completely changed his life, the 1831 upraising against the rule of the Holy See. After the bloody Restoration by the Austrian troops he was jailed for four months in Venice, and after he was freed he went in exile to Paris, where he was able to work in Gay-Lussac's laboratory. In 1833 he began to publish in French chemical journals, in 1840 he took French citizenship, and in 1842 he became professor of chemistry at the University of Rennes, where he remained until his death.[62] Malaguti was an excellent researcher and worked in several advanced fields of chemistry. His colleague and long-time co-worker A.E.Baudrimont thought that he was "the prince of the Italian chemists",[63] but it is probably more appropriate to think of him as a good Italian patriot and a good French chemist.

The reasons for including Jakob Moleschott (1822-1893) and Anton F. Dohrn (1840-1909) in this small review of 'nationality' problems will shortly be evident. Moleschott was born in Bois-Le-Duc in Holland, a physician of European fame and a hard-line materialist. When he was professor in Zurich he met the Italian historian of literature Francesco De Sanctis (1817-1883), who was a professor at the same Polytechnic. De Sanctis, a Bourbon subject, had suffered three years in jail between 1850 and 1853 and was afterwards banished; in 1859 he returned to Italy, and in 1861 he was appointed Minister for the Public Education in the last Cabinet of Count Camillo Cavour (1810-1861). In this post he called Moleschott to the Chair of Physiology of the University of Turin, where his influence on the medical studies was enormous.[64] In due time Moleschott was called on the Chair of Human Physiology at the University of Rome, became an Italian citizen and was named Senator of the Kingdom of Italy by the King. Anton Dohrn was born in Stettin (the Polish Szczecin); after long *Wander-jahren*, a degree in zoology, and other scientific peregrinations, he decided that the solution to the many problems of evolutionary theory was to be found in studying living organisms, especially those living in the sea. In short, in 1872, with the help of the local authorities, he founded the Naples Zoological Station, which became an extraordinary centre of research: extraordinary for its freedom, for the number of researchers who passed through the institution, and for its international resonance.[65] For historians, Dohrn is Italian by definition, as is Boscovich; both of their biographies are published in the *Dizionario Biografico degli Italiani*.[66]

At this point, after these diverse stories, we may look again at the use of the phrase 'Italian scientist'. When we refer to the times of Boscovich and Lagrange it may be used conventionally, as Boscovich himself suggested: it was only because of his long stay in Italy (or in the Italian States) that he might be said to be *en quel-*

que sorte Italian. During the lives of Avogadro and Malaguti, the definition takes on a neater contour - in the 1840's thousands of self-defined Italian scientists would gather in crowded meetings (see below). But immediately after Unification we find that scientists coming from Central and Northern Europe are considered Italians. From the point of view of so-called nationality, my sketchy biographical research has achieved nothing except to show how capricious a legal definition may be. In contrast, from the point of view of scientific relationships between the many European places where scientists have lived and done research, the small collection of pocket biographies reveals a clear pattern: for *cultural* reasons too, but essentially for *political* reasons, the first four scientists were involved in French science: Boscovich and Lagrange chose to go to Paris; for Avogadro there was no problem, France reached out for him; Malaguti, again, chose to go into exile to Paris. After 1859, the scientific policy of the liberal Governments and the hopeful enthusiasm for Unity stopped the forced migration of scientists; moreover, foreign researchers were called to Italian Universities or freely chose the Italian Nation as the seat of new Institutions. Behind all these movements of people and shifts of State borders, a glimpse of a more complex history can be seen. In the next sections, I will consider some aspects of the relationship between political power and scientific culture in Italy in the period considered here.

1. SCIENCE, CULTURE AND POLITICAL POWER

This part of my inquiry will be divided into three sections of unequal length; the first roughly corresponds to the period between the Peace of Aachen and the attack on Piedmont by the *Armée d'Italie*; the second to the Napoleonic turbulence; the third, longer, to the period between the Restoration and the Unification of Italy. I will pay more attention to the latter because the 'local' historical processes, albeit strictly connected with the frequent Europe-wide convulsions, ended in a unique result in the history of the Peninsula: its Unity.

1.1 An Age of Reforms

Ludovico Antonio Muratori (1672-1750) immediately perceived the Peace of Aachen as a fundamental event in the history of Europe. This priest, an erudite and a great historian, in the last year recorded in his *Annali d'Italia* (1749), wrote: "memorable must this peace be called because not only does it spread over the whole of Europe, but it is joined by the totality (*universale*) of the whole Earth".[67] The date of the Peace of Aachen is chosen conventionally as the beginning of an age of reforms introduced by the rulers of the Italian States, with rulers who were not Italian but Austrian in the vanguard. However, it is obvious that the winds of enlightenment had also been blowing for years in Italy. The very famous book by the Venetian Francesco Algarotti (1712-1764), entitled: *Il Newtonianismo per le Dame, ovvero Dialoghi sopra la Luce e i Colori* [Newtonianism for Ladies, or Dialogues on Light and Colours] is an example of this new attitude. Algarotti's life could be chosen as a paradigm of the life of a cosmopolitan man of letters. Born in a rich family of merchants, at fourteen he went to Bologna, where he studied philosophy and physical sciences; in the following years, in the Istituto delle Scienze, he performed experiences on optics, and enjoyed an intense relationship with the *letterati* of his *alma mater*. When, in 1734, he went to Paris he was already working on the *Newtonianismo*; there he became friendly with Maupertuis and Fontenelle, and to the latter he dedicated the first editions of his major work. Algarotti's book was printed for the first time in Milan, in 1737, without the *imprimatur* of the ecclesiastic authorities, and with a false indication (Naples) of where it had been printed.[68] This was not only the start of a triumphal journey of Algarotti's work through XVII century culture, but also an early start - a year before Voltaire - in the diffusion of Newtonianism in Europe.[69] All over Europe, enlightened ladies could read no less than 31 editions of *Newtonianismo per le Dame*, which had been translated in English, French, German, Dutch, Swedish and Portuguese.

Algarotti's work was far from being a mere popularisation. In the European context, the comparison with Voltaire's *Éléments de la philosophie de Newton* (1738) became obvious; in 1739 Eustachio Manfredi (1674-1739), mathematician and astronomer, wrote to his pupil Algarotti that the current judgement - in Bologna - on the two works was: "In his book Voltaire exhibits the geometry that he does not have, while Algarotti hides the geometry that he has".[70] But what interested the Holy Inquisition was not the (hidden) mathematical content of Algarotti's *Newtonianismo*, but its (hidden) Lockeian background, which induced the Roman Congregation to condemn the book, and to include it in the *Index librorum prohibitorum*.[71] The penetration of Locke's thought in Italy was one of the principal effects (and in its turn a cause) of the new opening of Italian culture to Europe, an opening "through the abysmal distances dug by the Counter-reformation."[72] Locke's *Essay* had been condemned in June 1734, and in November 1751 the same condemnation struck at Montesquieu's *Esprit des lois*, which had had a tremendous reception in Italy. In February 1766, the Santo Uffizio condemned the major work of Cesare Beccaria (1738-1794), *Dei delitti e delle pene* [On Crimes and Punishments], first published anonymously in Leghorn in the summer of 1764, and immediately famous all over Europe.[73] It is clear that, in the generation between Algarotti's book and Beccaria's Italian culture had joined in European efforts to create a more open society. At this point, it should be recalled that the phrase 'Italian culture' include many members of the Catholic clergy, who were integrated into the crowd of letterati living in Italy. It is certain that the official censure of Montesquieu's *Esprit des lois* and of Beccaria's *Dei delitti e delle pene* remained completely isolated from the growing disgust of public opinion at any despotic policy.[74]

In fact, attempts by the Holy See to exercise ideological control were somewhat erratic. This is particularly evident in the long polemic that saw Lazzaro Spallanzani (1729-1799) opposed to John Turberville Needham. The contention was about so-called spontaneous generation, and debate had been raging for years on the European scientific stage between Needham, backed by the famous Buffon, on one side, and Charles Bonnet, the 'preformist' naturalist of Geneva, supported by other lesser authors, on the other. The "Irish Jesuit"[75] had the merit of transforming the polemic from a wholly theoretical dispute to a confrontation with experimental results. Spallanzani, who had been ordained priest of a small local congregation in Modena, stepped into the controversy in 1765 with an essay based on accurate and original experiments (*Saggio di osservazioni microscopiche concernenti il sistema della generazione de' signori di Needham, e Buffon*);[76] Spallanzani sent the book to many well known letterati, including Voltaire, the Swiss Albrecht von Haller, and Bonnet, and immediately gained European fame, which, in its turn, induced Maria Theresa, Archdukes of Austria, to appoint him professor of natural history in Pavia (1769). Curiously enough, Needham had answered with a series of notes to the French translation of Spallanzani's Saggio, published in 1768 with a new, less challenging title *(Nouvelles recherches sur les découvertes microscopiques et la génération des corps organisés)*. This polemic ended in 1776, when Spallanzani published *Opuscoli di fisica animale e vegetabile*, a work that compelled Needham to retire from the controversy. It has been remarked that both protagonists were Catholic priests;[77] and as far as I know, the Holy Inquisition did not go into the question, although was ideologically sensitive from several points of view.[78] In 1768, another long and lively polemic, which involved scholarly Europe, had been kindled by Spallanzani in *Prodromo di un'opera da imprimersi sopra le riproduzioni animali*, again published in Modena, but immediately translated into French and published in Geneva (*Programme, ou Précis d'un ouvrage sur les reproductions animales*, 1768). Here, *riproduzioni animali* meant the regrowth of mutilated organs in animals, a theme that impassioned the European illuminists, and that, under the thrust of Spallanzani's work, became so popular as to be covered in the newspapers.[79] Spallanzani's long scientific career demonstrated that it was perfectly possible to found a new experimental *practice* - applied to *living* beings - while working in Italy, publishing in Italian, and being a Catholic priest.

I have emphasised the word 'practice' because the Church was more attentive to the visible philosophical (or ideological) consequences of science than to the epistemic content of the research. If science remained a collection of 'facts', or 'a method' without trespassing on the epistemological limit of a 'local' truth, there was no necessity for intervention.[80] In regards to this attitude, what happened to

Boscovich during his stay in Rome in the years 1744-1757 is emblematic. These were the years during which the Jesuit constructed his European fame as a mathematician, and they were also the years in which he - among other activities - deepened his understanding of Newton's works, in studies that later led to the writing of his masterpiece, the famous *Theoria*. In 1748, two lectures with Boscovich's reflections on Newton's optics and a theory on the nature of light were presented to the Jesuit power-centre, the Collegio Romano; the lectures were published in two parts with the titles *Dissertazionis de Lumine pars prima[pars secunda] publice propugnata in Seminario Romano Soc. Jesu*. The first part was a critical examination of the *Axiomata* and the *Definitiones* that give the rules of the physical behaviour of the *lumen* in Newton's *Optice*; the second part is on the physical structure of the matter of light and gives the first draft of the field-like theory of the physical world that he was to complete and extend in *Theoria*.[81] This direct dialogue with the spirit of Newton wounded the orthodoxy of the Jesuit Order so that, despite the protection of the Pope Benedict XIV, the Order reacted resentfully to the "excess of liberty and of originality of its member",[82] and in 1759 Boscovich left Rome and his Chair at the Collegio Romano; in 1763 he would accept a chair at the University of Pavia.

It is time to consider this University, the most important scientific centre in XVII-century Italy. Pavia's standing as a scientific centre of international repute lasted beyond the fall of Napoleon, but the oppressive cultural milieu of the Restoration almost extinguished this excellent tradition. In the half century considered in this section of the paper, however, Pavia was always under the rule of Vienna, and its University has to be considered in the broad context of the Austrian Empire. In this connection at least three figures must be mentioned, at three different levels of the political and cultural life of Lombardy. The person responsible for the policy of the Italian provinces of the Empire was Wenzel Anton Kaunitz-Rietberg, the statesman who directed Austrian foreign policy from 1753 until 1792. The Lombard territory was not geographically connected with the Austrian territory in Süd Tyrol, and depended directly on the Vienna Chancellery. Seen from Vienna, Lombardy might well look tiny compared with the enormous extent of the Empire; Kaunitz had been known to say that when he thought about Italian affairs (and Dutch ones too) was in the morning, while he was putting on his stockings. Perhaps it was not quite so simple, certainly Lombardy, and Pavia University, was considered an outpost of the Austrian monarchy.[83] The right man for this frontier policy was the Count Carlo Firmian (1718-1782),[84] the Plenipotentiary Minister of the Austrian Chancellery in Milan, who encouraged the work of many *letterati*, who were interested in the scientific and technological development of Lombardy. It is at this third level that we find intellectuals like Paolo Frisi (1728-1784), a member of the Order of Barnabiti; at only 25 years old, he had been elected a member of the Paris Académie des Sciences, and for many years he studied the form and structure of the Earth. A *letterato* of European renown, he actively participated in debates and in proposals about the reform of the University of Pavia and the establishment of the Palatine Schools in Milan; he contributed to literary journals edited in Milan by Lombardy's most advanced illuminists, and at the same time he studied the control and rationalisation of the course of rivers and canals in Lombardy.[85] Notwithstanding the cultural stature of men such as Frisi, the Milanese proposals for reform were judged by the Vienna court to be backward;[86] in 1779 Kaunitz indicated the University of Göttingen as a model. Vienna nevertheless had appointed many significant researchers to important teaching posts in Pavia, including Boscovich in 1763, for mathematics; the anatomist Pietro Moscati (1739-1824), also in 1763; Gregorio Fontana (1735-1803), in 1764 for logic and metaphysics, and after 1768 for mathematics; Spallanzani in 1769 for natural history.

By the way, the appointment of Boscovich in Pavia had a particular meaning in the context of the renewal of the professional instruction for civilian engineers. In the same year 1763 Frisi was called from Pisa, and appointed to the teaching of Mechanics and Science of Waters (*Scienza delle Acque*) at the Palatine Schools in Milan. Both the appointments had the meaning of making public and compulsory for the future engineers a teaching which, so far, had been private and optional, but the treatment of the two mathematicians in canonical was very different. Boscovich, the 'foreign' Jesuit, had an exceptional salary of 4500 lire per annum, while Frisi, the Lombard Barnabita, had only 2000 lire. Moreover the complete independence of Boscovich from the lay authority was stressed by the fact that his salary was not bound (*legato*) to the purchase of books or instruments. The Jesuit mathematician lived on the income of the Collegio of the Society of Jesus in Pavia, and in a letter to Firmian (7 November 1766) he openly stated that he received orders only "by his Sovereign, who was the Pope".[87]

The list of scientists called to Pavia lengthened as time went on: between 1776 and 1796 Chairs were given to Giovanni Antonio Scopoli (1723-1788), for chemistry and botany; Alessandro Volta (1745-1827) for experimental physics; Antonio Scarpa (1752-1832), for anatomy and surgery, Lorenzo Mascheroni (1750-1801) for algebra and geometry; Giovanni Rasori (1766-1837) for medical pathology; Luigi Valentino Brugnatelli (1761-1818), for chemistry.[88] Many of these names will crop up again as being involved in later scientific or political events (Rasori and Brugnatelli were appointed in the crucial year of the Napoleonic invasion). My point here is just that distinguished scientists, as well as funds for research and teaching structures, arrived in Pavia. Starting from 1771, the University buildings were extended, and a new chemical laboratory was constructed; and on Scarpa's initiative a splendid new anatomy theatre, with annexed laboratories, was inaugurated in 1785. At about the same time, the botanical gardens were transferred to a more suitable place, and a chemical laboratory was added by Scopoli; in 1771, to Spallanzani's delight, the first collections for the Museum of natural history arrived in Pavia from Vienna; finally in 1787 the magnificent new physics theatre was inaugurated for Volta.[89] It is also worth noting that both of these fine amphitheatres were designed by the neo-classical Austrian architect Leopold Pollack; essentially he was active in Lombardy and was one of the Lombard aristocracy's favourite architects. However, the administrative dependence of Pavia's building program on Vienna is confirmed by the presence of the drawings are kept in the Vienna State archives.[90]

In Piedmont, the reform of the University began very early, in the 1720s, but the results were less spectacular than in Pavia. One scientist, at least, must be mentioned: Father Giovanni Battista Beccaria (1716-1781); and one research structure stands out: the laboratory of hydraulics at Parella, near Turin. Beccaria, a member of the Order of the *Scolopi*, was appointed to the Turin Chair of physics in 1748, where he introduced the experimental method, in contrast with the Cartesianism of his predecessors. In 1753, he published an important contribution to the study of electricity, *Elettricismo artificiale e naturale libri due*, which was translated into English thanks to Franklin's interest. Beccaria corresponded with many European scientists, and in 1757 he was elected Member of the Royal Society. In 1759, Boscovich persuaded the King of Sardinia of the prestige deriving from the measurement of an arc of meridian, and Beccaria was given the task;[91] he was helped in this by his University assistant, Domenico Canonica (1739-1770). The results were somewhat surprising when compared with barely a dozen of similar measures available at the time, because of the anomalies introduced by the presence of the Alps;[92] however they where published under the title *Gradus taurinensis* (Turin, 1774). The very nature of the measures made by Beccaria and Canonica attracted the keen attention of the learned world, and the 'Turin anomaly' kindled a sharp dispute with the astronomers of Paris. Plana would later demonstrate that Beccaria was right (see below).[93] In this stimulating milieu Beccaria established the first school of physics active in Italy in the second part of the XVIII century: his students included Lagrange, Cigna, Saluzzo, and the successor to the Chair of physics, Vassalli Eandi; moreover, his researches inspired the physicists of Pavia, Volta and Barletti.[94]

From the 1760s onwards Turin had an important scientific and technological structure, the *Stabilimento delle esperienze idrauliche*, at Parella, designed by the University professor of hydraulics, Francesco Domenico Michelotti (1710-1777). The center of the installation was a tower[95] seven meters high that permitted observations and measures on several phenomena interesting for a rational control of irrigation. The laboratory gained European fame, and was the best result of the 'practical' intentions of the Piedmontese reformers; it later became the experimental site of a famous Italian hydraulic engineer and mathematician, Giorgio Bidone (1781-1839), who confirmed the exceptional utility of the large plant at Parella.[96]

The reforms also reached the lesser Italian States. In the Duchy of Parma, after an Observatory (Specola) had been established in 1759, there was a progressive development of the scientific institutions, with the foundation of the Museum of Natural History (1766), the Botanical Garden (1768), the Physics Cabinet (1770) and the Chemistry Cabinet (1771).[97] At that time, Parma was a centre from which French culture was spreading into Italy, largely owing to the presence of Du Tillot, the Duke's French Prime Minister, and of the philosopher Condillac.[98]

An important sign of the rulers of the Italian States' changing attitude towards science was the foundation of the public observatories. The first Italian public observatory was established in Bologna, at the end of a process that had started at the beginning of the XVIII century when the Count Luigi Ferdinando Marsigli (1658--1730) charged Manfredi and Vittorio Stancari (1678-1709) to construct a Specola. In 1712, when Marsigli presented his books, collections and instruments to the local Academy, the *Constitutiones* of a new scientific institution on the model of the French Académie des Sciences were approved by the University Senate, and a few years later the Istituto delle Scienze, an institution of momentous importance in Italian science of that century was officially opened (March 1714).[99] The new observatory was opened in 1727, and was also equipped with new instrumentation, designed by Stancari. It is noteworthy that Bologna Observatory was the fifth public observatory in Europe after those of Paris, Greenwich, Leiden and Nurimberg.[100] The second Italian public observatory State was established in Pisa, in the Grand-Duchy of Tuscany; regular observation only began in 1765 with the second Director, Giuseppe Antonio Slop de Cadenberg (1740-1808). The Grand-Duke intended to have another Specola in Florence, and the task of building it was given to a capable scientist, Felice Fontana (1730-1805), who among other abilities, was also a good instrument technician. However, the Florence observatory, begun in 1775, was not yet in working order at the end of the century.

Cases such as this show that the will of an 'enlightened' prince was not enough to produce a healthy scientific structure.[101] A 'surreal' example, which confirms the arbitrary nature of the political power's intervention in the scientific 'reforms', is the appointment of Ranieri Gerbi, (1763-1839). He took his degree in medicine in 1789, upon which the Grand-Duke appointed him to teach mathematics in Pisa, and from 1797 he held the Chair of 'theoretical physics' in that University. The chronicle of the time depicts the 'physicist' as keen on entomology, and affirms that his most important discovery was an insect, named *curculione antidontalgico*, useful against toothache (*Storia naturale di un nuovo insetto*, 1794).[102]

The story of the foundation of the famous Observatory of Brera was completely different. In 1764, Boscovich, was had been appointed to the Chair of mathematics in Pavia in 1763, decided to pass the following summer vacation in the Jesuit College in Milan, in the grand Palazzo Brera. There he was asked by the Dean of the College to design a new observatory; Boscovich agreed, his plans were approved by Count Firmian, by the Viennese court and by the Jesuit Order. With lightning efficiency, the Observatory tower was ready in 1765, and in just a few years it was fitted with instruments. The funds were donated by the Order, and personally by several Jesuits, including Boscovich.[103] Modern astronomy also reached Rome, but not with a public observatory. In 1777, the Duke of Sermoneta, Francesco Caetani (1738-1808), commissioned Giovanni Battista Audifredi (1714-1794) to design and build an observatory in his Roman palace. This first, private observatory took the name of its founder (Specola Caetani) and had good instrumentation. In 1787, - when the Jesuits were disbanded - Giuseppe Calandrelli (1749-1827), a Canon, constructed an astronomical observatory, with modest instruments, in the Collegio Romano; the researchers of the two Roman institutions were always rivals.[104]

1.2 Two passionate decades: Napoleon's invasions and wars

In 1791, in the *Commentarii* of his University, the Bolognese anatomist Luigi Galvani (1737-1798) published an essay that started one of the most famous polemics in the history of science.[105] The memoir was the previously unpublished result of a decade of work on animal electricity,[106] and was rigorously written in Latin as were all Galvani's signed works. Alessandro Volta was immediately involved in verifying Galvani's theses by a colleague at Pavia, the priest Bassiano Carminati (1760-1830). After a short period of almost complete agreement between Pavia and Bologna, the controversy exploded that was to open the way to Volta's momentous invention of the battery.[107] As is well known, Volta communicated his extraordinary invention to Joseph Banks in a letter dated 20 March 1800, which was read before an amazed Royal Society, and published in the *Philosophical Transactions*; but, by then, Galvani had died in despair and distress, because he had refused an oath of loyalty to the new-born Cisalpine Republic.[108] Italy was quickly becoming a French dependency: on 14 June 1800, Napoleon won a decisive battle at Marengo, in Piedmont, and for the following fifteen years it was he who decided the fate of Italians.

In April 1796, the French troops lead by Napoleon had defeated the Austrian and the Piedmontese armies. Thus began a long - and intoxicating - tragedy for the Italians that ended with the unbelievable backlash of the Restoration.[109] It is possible to see some constants in the relationships between science and political power during these years. To be sure, the Napoleonic period falls into three phases: the so-called 'Jacobin triennium', 1796-1799, characterised by Republics all over Italy; the very short return to power of the ruling dynasties, backed by the Austrians and the English, and; the years from Marengo until the Restoration. The contrast between the three Jacobin years and the fifteen years of direct French domination is clear-cut, because the former were hectic years of hope, and of terrible errors committed by the Italian Jacobins, while in the latter period the Italian stage had a single protagonist, Napoleon. The period 1800-1814 was an epoch of rationalization and social conservatism, in which the borders between the Italian States changed more frequently than in any other period since the XV century.[110] In this situation, many scientists chose to be quiet, other were actively and dramatically involved in the political convulsions, and still others were able to come unharmed through all political changes. This last was the case of the greatest scientific celebrity of the time, Volta, who was an Austrian subject for 51 years, then a citizen of the Napoleonic Cisalpine Republic and Kingdom of Italy for 18 years, and still later an Austrian subject for 13 years: under each regime he was always "prudent, devoted, faithful and obsequious".[111] Volta's political indifference may be contrasted with the commitment of other professors of the University of Pavia, such as Moscati and Rasori. Moscati, who was older than Volta, was a representative of the Lombard enlightenment and participated in the Directory of the Cisalpine Republic, concerned in particular with the reform of public health. Rasori, a pupil of Spallanzani, was a generation younger than Volta and an active Jacobin. He was scientifically oriented towards British culture; he translated Erasmus Darwin's *Zoonomia* into Italian, and adopted in his teaching the medical theories of the Scottish physician and biologist John Brown.[112] In 1797, still very young, Rasori was named Dean of the University, to the great envy of his senior colleagues, and after the arrival of the Austrian and Russian troops in Milan he left the city and joined the French Army in Liguria. Other professors were captured and jailed by the Austrians; among them Moscati, the physicist Carlo Barletti (1735-1800) who had been among the founders of Lorgna's Società Italiana (see below), the mathematician Gregorio Fontana, who had, since 1797, been on police files as "covert Jacobin and atheist", and the professor of pharmaceutical chemistry Francesco Nocetti, who died in the course of deportation to Dalmatia.[113]

The Tuscan, Neapolitan and Venetian reformers were less fortunate than those of Lombardy. In Tuscany, after the defeat of the French troops in 1799, bands of peasants organised by aristocrats and clergymen in the district of Arezzo raged to the cry *Viva Maria, viva l'Austria, abbasso l'albero della libertà* [Long live Maria, long live Austria, down with the tree of liberty]. In Florence, as in the rest of Italy the scientific community was split; for example Giovanni Fabbroni (1752-1822), chemist, naturalist and pupil of Fontana's, was very cautious on the occasion (he was later able to survive any political change[114] - like Volta). When the bands arrived in the city of Florence, many 'suspects' were arrested, among them, perhaps at the instigation of Fabbroni, Felice Fontana, the founder and keeper of the famous Florentine Museo di Fisica, and a member of the liberal wing. Despite his age - he was 69 - he was manhandled by the populace, and jailed for a while. The sectarian nature of the movement is exemplified by the attacks on the Jewish communities; on 28 June the bands from Arezzo entered Siena and a "bestial massacre" of Jews followed.[115] In Naples too, as in other Italian States, the part of the learned society most closely involved in the so-called Jacobin policy was the community of physicians. Domenico Cirillo, a physician and botanist, was engaged in the reform of the jails, true hells for criminals as well as for "many workers unjustly put in chains"[116]. After the defeat of the French, the Neapolitan Republic was daringly defended by a 'sacred battalion' (*battaglione sacro*) of young physicians, and after the fall of the Republic, betrayed like the other Jacobins by Admiral Nelson, Cirillo was sentenced to death and hanged.[117]

The Venetian Jacobins were also betrayed, but by Napoleon. Vincenzo Dandolo (1758-1819) was a protagonist and a witness of the disgraceful story that ended with the consignment of the glorious *Serenissima* Republic of Venice into the hands of the Austrians. Dandolo had studied in Padua, and was a pupil of Count Marco Carburi (1731-1808), professor of Chemistry at Padua University since 1760 (the University's first). By profession, Dandolo was a pharmacist (or better a *speziale*) and rich manufacturer of drugs, but his importance in the Italian community of chemists came from his translations of many important French texts of the time, including Lavoisier's *Traité élémentaire de chimie*, published in 1789 and translated in 1791. Dandolo and Carburi were on opposite sides over the reform of chemistry, but they equally sought a democratic reform of the oligarchic Republic: in 1796 the pro-French factions met secretly in Carburi's home in Padua and in back rooms of Dandolo's shop in Venice, while waiting for Bonaparte's troops.[118] At last the French arrived, and Dandolo became a member of the new Local Government of Venice, but Ugo Foscolo, then secretary of Venice *Municipalità*, informed the Venetian citizens on 19 October 1797 that two days earlier Bonaparte had signed a truce with the Austrians. Dandolo and three other delegates tried to go to Paris to prevent the demise of the Republic, but Bonaparte intercepted them, and, notwithstanding an eloquent appeal from Dandolo, remained inflexible in his decision to give up Venice to Austria.[119] It was an act worthy of a dynastic prince of the XVIII century: in the style of "arithmetic diplomacy",[120] Napoleon ended the independence of the ancient Italian Republic. Before leaving Venice, the French burned the *Bucintoro*, the magnificent ceremonial ship of the Doge, pillaged the Arsenal, and sent the four famous bronze horses of St. Mark's to Paris.

Many of these infamous acts of war were carefully planned. In 1796 the *Directoire* in Paris had named a special commission to choose the masterpieces and cultural treasures to be brought to France: its members were the chemist Berthollet, the geometrician Monge, the naturalist Thouin, the sculptor Moitte, and the painter Barthélemy. Bonaparte also personally chose some plunder, including Fontana's anatomical wax collection in Florence. On 19 February 1797, Pope Pius VI signed a treaty in which the provinces of the Papal States in the North, including Bologna and Ferrara, were 'ceded' to France. A clause of the treaty also 'ceded' a great number of works of art; they were transferred from Rome to Paris in long caravans of heavy wagons.[121] The worst aspect of this sad story is that some scientists, who like Spallanzani had been personally unharmed proved very 'sympathetic' to this pillage,[122] and other intellectuals agreed, as did the celebrated musician Luigi Cherubini (1760-1842) who *advised* the transfer to Paris of the precious collections of music kept in Bologna.[123]

At the beginning of 1801, Bonaparte 'stabilized' the situation in Italy by a shrewd political manoeuvre. He summoned an assembly of 450 notables to Lyons (*Comizi di Lione*), including University professors and representatives of scientific organizations. The assembly had the task of giving the Italian Republic a constitution, but it was only "the parody of a constituent assembly".[124] (The Republic would clearly going to be presided over by Bonaparte himself). Among the scientists were Brugnatelli and Volta from Pavia;[125] the Società Italiana delle Scienze was represented by two members, while its President, the astronomer Antonio Cagnoli (1743-1816) represented the Military School of Modena; the mathematician Luigi Rangoni (1775-1844) was a representative of the 'property-owners' (*possidenti*).[126] In regard to the question of the organization of science the *Comizi* decided to establish a National Institute of the Italian Republic in Bologna; the explicit model was the French Insitut National des Sciences et des Arts, and its task was to be "the collection of the discoveries and the perfecting of sciences and arts", with the intention of directing scientists towards the practical interests of society. In reality the National Institute followed the destiny of its French counter-part and became a technical organism of the State.[127] On the whole, the political regime that was sanctioned by the *Comizi* was very conservative, as were its members, and in the following years any hope of independence for the Country disappeared.

The dependence on France was not only political. The wars in the title of this section are not only the wars fought in Italy by French, Austrians, Russians and so on, but essentially wars fought by Italian soldiers everywhere in Europe. Hundreds of thousands of young Italians were recruited into the Napoleonic Armies[128]; they were sent to fight and to die for causes that had no relationship with their Country, and the impact on collective feelings towards France was lasting. In 1844 Ignazio Cantù published a book on *L'Italia scientifica contemporanea*, consisting essentially of a collection of short biographies of Italian men of science; its immediate target was the sixth Meeting of Italian physicians, naturalists and scientists, which was about to be held in Milan. In two of these biographies, Cantù found ways of speaking about Italian losses in the Napoleonic wars: in one, a physician is reported to have participated in the battle on the Beresina river, "where the flower of Italy was frozen (*ingelidito*) among the ice of Beresina and the icy winds of Moscow".[129] In the second, a high-ranking officer of the Engineering Corps is reported to have written a *Storia delle campagne e degli assedi degli Italiani in Ispania dal 1808 al 1815* [History of the campaigns and of sieges in Spain from 1808 to 1815]. Cantù draws up a bloody balance: out of 30,183 Italians who participated to the Spanish wars, only 8,958 returned to Italy, and 21,225 died in Spain.[130] It is not surprising that many eminent political writers of the Italian *Risorgimento* were against opposed to any emulation of French culture, in defence of a national character.[131]

1.3 The Risorgimento *and the Unity of Italy*

You do quite right to stipulate for a grand Laboratory there. A chemical Laboratory in Rome! is a glorious idea. Truly we live in an age of progress! I congratulate you most cordially on your invitation to put yourself at the head of such an important movement.

E.Frankland to S.Cannizzaro, on 3 December 1871[132]

Within the space of two years, 1859 and 1860, for the *first time* in her long history Italy became a State with territorial unity extending from Piedmont to Sicily.[133] This event was the unexpected, and unforeseeable, result of an extraordinary and swift convergence of three different processes. The dynastic war of the Kingdom of Sardinia and of the French Empire had won Milan and Venice from Austria in 1859, and at last the influence of Vienna in Italian affairs came to an end. The overthrow of the of Bourbon rule in the Kingdom of the Two Sicilies was the spectacular outcome of the military talent of Giuseppe Garibaldi, and of the boldness of his Red Shirts and of the other thousands of volunteers who increasingly relied to the expedition. A third important process was made up of political demonstrations, popular riots and armed uprisings that in various ways freed Tuscany, Bologna, Modena and Parma from their rulers. Naturally, these three processes were, in their turn, the result of the decades of efforts that are summed up in the word *Risorgimento*: the bravery of the Italian patriots, the long-lasting philosophical, cultural and political discussion among the intellectuals, the slow but visible development of a capitalist economy in the North and in Tuscany, and, finally, the colliding interests of the reigning dynasties, 'local' and European. To all these components Italian scientists contributed with mixed and sometimes contrasting feelings.

As is well known, in the period from the Restoration to the proclamation of the German Empire, the European bourgeoisie fought both openly and covertly to gain political power - with its nationalistic content. In the Italian States, as in the German ones, the issue was complicated by the lack of territorial political unity, which appeared to be a necessary, ideological condition for a national State. The Continental political context was everywhere so similar that a 'revolution' in one country had immediate resonance in other parts of Europe; in the description that follows this 'domino' effect will be drastically condensed, and the real process of transmission of 'historical momentum' will remain out of sight.

Before the eventful two years of 1859-60, the situation in Italy was such that there was either a major uprising or a war in each of three decades: 1820-21, 1831, and 1848-49. It can well be imagined that scientific life, still depressed by the conservative turn of the Restoration, was in bad condition. It was in *really* very bad condition at least until about 1840, when in several Italian States there were indecisive attempts to improve it.

I will consider this long and troubled period, by looking at each of the most significant events of the *Risorgimento* from a particular vantage point, where the events in question had the biggest effect on the local scientific community - beginning from the riots in Turin in January 1821.

The uprising in Spain in January 1820 was followed in July by a military revolt in the Kingdom of the Two Sicilies. Unfortunately for the liberal movement, the two parts of the Kingdom showed themselves to be really 'Two Sicilies': the new - provisional - rulers of Naples and Palermo soon began to quarrel angrily about the independence of the Island from the 'continent'. The quarrel had heavy consequences for the whole of the *Risorgimento:* the liberal Government of Naples sent troops against the liberal Government of Palermo, and the revolt was repressed with much boodshed. In Piedmont, the Neapolitan events were followed with concern by patriots, and by the Turin police, which arrested several students for trifling reasons on 11 January. The consequent riot of students and some army detachments which followed was easily repressed by Austrian troops (8 April 1821); meanwhile, the Neapolitan rebels had also been crushed by the Austrians at Rieti (7 March). In Turin, 223 students were persecuted, and many were definitively banned from the University; a student of chemistry was sentenced to death *in absentia*. The effects on University life were of extreme importance because teaching was dispersed throughout the provinces to avoid crowding of students into the capital; (this highly anomalous situation lasted until 1842).[134] The fate of Avogadro's project for an advanced experimental physics laboratory is emblematic of the effects on scientific activity at Turin University. In 1819 Count Prospero Balbo (1762-1837) had contacted Avogadro for a Chair in theoretical physics (*fisica sublime*), and in 1820 the Turinese physicist was appointed Professor in the University of his own city. The laboratory project included teaching and research, and Avogadro had the first innovative instruments made when the Chair was suppressed by Decree on 31 July 1822. Avogadro had only been sympathetic with the student movement, and probably he only lost the Chair because of his friendship with the liberal family of Count Prospero Balbo, who had been instrumental in his appointment. Avogadro only returned to his academic post in 1834.[135]

The second wave of revolts occurred over the years 1830-31. Several Italian States were involved, but the most serious events occurred in the Duchies of Parma and Modena, and in the Papal States. In the case of Parma, the important role of the physicist Macedonio Melloni (1798-1854) played an outstanding role. He had studied in Paris and had initiated a productive collaboration with the Florentine physicist Leopoldo Nobili (1784-1835), an accurate experimentalist who, in 1825, had constructed the first astatic galvanometer, and had used it as a sensitive instrument to study the electric currents associated with muscular contraction.[136] In 1829 Nobili had told Melloni that he had constructed a 'thermo-multiplier', a sensitive thermo-electrical battery connected with his galvanometer, and Melloni adapted the new detector for the study of radiant heat. The collaboration was yielding very good results when grave political events interrupted it. On 15 November 1830, in the inaugural lecture of his course at the University of Parma, Melloni exalted the behaviour of the Parisian students in the riots of the past July. Melloni's students carried their professor shoulder-high, but the following day, he was dismissed and banished from Parma. After a short stay in Paris, Melloni returned to Parma and sat in the Provisional Government until March 1831, when the Austrian troops compelled him and other patriots to go into exile. Nobili, too, suffered for his part in the general turmoil, as well as the young pharmacist from Bologna Faustino Malaguti, whose status as an Italian scientist was discussed above; Malaguti never returned to Italy.

Melloni, however, returned to Italy in 1837, thanks to pressure exercised by Dominique François Arago and Alexander von Humboldt on Metternich, and by Metternich on the gracious Duchess Maria Luigia of Parma. Melloni had studied the infrared spectrum with new optical instruments; on the base of his experimental results, he had argued the wave-wave-like, light-like nature of radiant heat and, to explain the results, he proposed the use of the wave theory of light before the Académie des Sciences in Paris. The French scientists were unclear in their response, but Faraday was extremely interested in Melloni's papers, and in 1834 he repeated Melloni's experiments before the Royal Society, which awarded the Italian physicist the Rumford Medal. In 1835, Melloni presented a new *mémoire* to the Académie; this time, an exacting commission was set up, composed of Poisson, Arago and Biot, [137] and their positive judgement sanctioned Melloni's scientific fame.

After the bloody and cruel[138] repression of the riots, the police remained very attentive to any sign of conspiracy, but they were quite unable to distinguish between the true and false enemies of the ruling dynasties. As in many other regimes, intellectuals were the best candidates for the police records, and this kind of political attention to scientists' activity did not help their independence of mind. In some cases, repression struck very young people, such as Francesco Colombani (1813-

1865) who in 1833 was suspected of affiliation with the liberal movement and compelled to flee abroad. Like many others before him, he went to Paris, where he was admitted to the Ecole des Ponts et Chausées; there he was a pupil of Gustave-Gustave-Gaspard de Coriolis, and as *ingénieur-élève* he worked with the engineer and physicist Emile Clapeyron. After an amnesty, in 1838 Colombani was allowed to return to Lombardy, and his activity as contributor to the most important technical and cultural journal of the period, the *Politecnico* of Carlo Cattaneo (1801-1869) stands out. The *Politecnico* was published for five years (1839-44) in the period dealt with here, and again (1859-65) after the annexation of Lombardy to Piedmont. Many representatives of the new Lombard bourgeoisie wrote for Cattaneo's journal: scientists, technicians, engineers and entrepreneurs.[139] The fact that a journal like the *Politecnico* could be published and had an audience was itself a sign that the times were changing.

I want now to examine the extent to which the political and cultural situation in Italy was polycentric, and in order to do so, it is necessary to look at some developments that were taking place in the relations between political power and science in the Papal States, in Tuscany and in Piedmont.

In the Papal States, the torpid - or morbid - condition of scientific teaching and research began to change, thanks to Cardinal Ercole Consalvi (1757-1824), who was Secretary of State to Pope Pius VII in the years 1816-1823.[140] In 1816, at the Sapienza University in Rome Consalvi established a new Chair for physics, significantly named *Fisica Sacra*. The Chair was given to Feliciano Scarpellini (1762-1840), who had been astronomer of the Specola Caetani. This was the spear-head of a cultural policy that set theological and confessional distortion of science. criteria firmly ahead of intellectual and epistemological rigor. The cultural programme to be pursued from the Chair was articulated in 1837 by the mathematician S.Proja thus: "This Chair intends to apply the natural sciences to the consideration of the supreme works of nature, with the dual aim of glorifying the name of this divine author and of refuting the errors that derive from the misuse (*abuso*) of the sciences themselves".[141] This manifesto was published in the *Giornale arcadico di scienze, lettere ed arti*, which had been established in 1819 and was for decades the principal scientific journal in Rome. The name of the journal, recalling the literary Arcadia founded in Rome in 1690, gave a clear signal that the cultural time in Rome was, at best, a century behind. That the city was in the rear-guard seems to be confirmed by the fact that, as late as 1841, the *Giornale arcadico* was against the great *Encyclopédie ou Dictionnaire raisonné des Sciences, des Arts et des Métiers*, which it defined as "a very powerful instrument of the universal subversion of ideas and of beliefs", and stated clearly that science should refrain from adhering to that subversion.[142]

While the times of the Eternal City were reluctantly evolving, in Tuscany there was a quicker awakening. In the early 1840s, Grand-Duke Leopoldo II appointed a formidable quartet to Chairs at Pisa: Ottaviano Fabrizio Mossotti (1791-1863), Leopoldo Pilla (1805-1848), Carlo Matteucci (1811-1868), and Raffaele Piria (1813-1865). It is doubtful that the Grand-Duke had much idea of the scientific and political potential of these four researchers. Mossotti was born in Novara (and thus was a subject of the Kingdom of Sardinia). He had been assistant in the Brera Observatory from 1813 until 1823; he was the best Italian theoretical physicist of the XIX century. In several memoirs, using brilliant mathematics, he proposed a very evocative theory of dielectric that moved within the mainstream of contemporary physics, while also trying to look from an unified point of view at the phenomena studied in optics, thermology and electrology. His work had a profound impact on Faraday, who presented the Italian physicist's results to the Royal Institution in 1836.[143] The second member of the quartet, Leopoldo Pilla, was born in a village in the district of Campobasso, (thus he was a subject of the Kingdom of the two Sicilies). He taught mineralogy in Naples; the institution of his Chair at Pisa was a scientific event, being the first Chair in geology in Italy.[144] Matteucci was born in Forlì (a subject of the Papal States) and took a degree in physics at the early age of seventeen, in Bologna. After a short stay in Paris, in contact with Arago and E.Becquerel, he returned to Italy, to Ravenna, where he worked in the laboratory of the local hospital, from 1836 to 1840.[145] In 1836 he sent the Académie des Sciences a note on *Expériences sur la Torpille*; this and his following papers, fully inside the Italian tradition of electro-physiology, brought Matteucci European fame and led to his invitation to Pisa, recommended by the ubiquitous and providential Alexander von Humboldt.[146] Last of the four, Piria was born in a village of the far South, Scilla (like Pilla, he was a subject of the Kingdom of the Two Sicilies). He had studied in Naples, obtaining a degree in medicine, and had then gone to France where, in Dumas' laboratory, he had obtained excellent results in the most advanced fields of organic chemistry. In 1839 he had returned to Naples, and, within a short time, Melloni - who was working in Naples - and Matteucci persuaded the Grand-Duke to call the young chemist to Pisa. There, over the following years, he succeeded in laying the foundations of the first Italian school of chemistry.[147]

While the reorganization of the University of Pisa was in progress, a group of naturalists and scientists, working in the medical and scientific institutions of the Grand-Duchy of Tuscany, proposed a meeting of Italian scientists. The model was that of the German Gesellschaft Deutscher Naturforscher und Aertze, and the first *Riunione* was actually held in Pisa in 1839. The series of these yearly meetings lasted until the revolutions, and the war, of 1848, and their significance for the scientific culture of the Risorgimento will be analysed in section 2.2 below. At this point, I will describe an incident in Matteucci's life that illustrates the difficult situation of liberal Roman Catholics in Italy. In 1835, in Forlì, Matteucci had published a monograph on "rational scientific method".[148] As I said, Forlì was in the Papal States, so the booklet had passed the censure of the Holy See, but when Matteucci tried to publish a second edition, the Inquisition summoned the now-renowned scientist to 'discuss' a few dangerous propositions, just at the time when the other researchers were going to Pisa for the first meeting. Probably the timing was not fortuitous;[149] what is certain is that the suspicions of the Inquisition concerned Matteucci's opinions about the physical basis of life. In his monograph he had argued in favour of a physicalist interpretation of life, claiming that it has passed "the test of experience", the "unique characteristic of truth." In the same passage he had written: "The great progress of our century is to have freed the study of natural sciences from the terrible prejudice that made them suspect, and almost incompatible with religious belief (*spirito religioso*)".[150] That proved to be wishful thinking: living in the middle of the XIX century, Matteucci escaped any physical consequences, but the second edition of his monograph was never published. The Roman Catholic Church has always appreciated the symbolism of events, and we cannot refrain from doing the same: in September 1839, while more than four hundred physicians, naturalists and scientists were going to Pisa, the splendid birthplace of Galileo, elected for the occasion as the centre of Italian science, one of Italy's best scientists was repeating the journey of the great Galileo to the centre of Roman Catholicism, for a discussion with the Holy Inquisition.

In Piedmont, after the riots of 1821, the cultural and political situation remained stagnant for many decades. The crowning of Carlo Alberto as King of Sardinia in 1831 kindled patriots' hopes, they were almost immediately extinguished in 1833 by the savage repression of a republican plot in the Army.[151] In 1835, Carlo Alberto began a very slow conversion to more liberal politicies, which became more visible in the early 1840s. One fact will illustrate the stultified situation of culture in Piedmont in the three decades following the Restoration: in 1842, when the King permitted the establishment of an agrarian association, it was the first case of an open, public society being permitted; what would have been a minor episode in the life of countries such as France or Britain was read as a sign of new hopeful development in Piedmont.[152] The effects on Piedmontese science of Carlo Alberto's very mild reforms were negligible, while at the political and cultural level a few books published by the moderate wing of the *Risorgimento* were much more relevant. In Brussels in 1843, Vincenzo Gioberti (1801-1852), a Turinese priest banished after the plot of 1833, published a book with the very significant title: *Del primato morale e civile degl'Italiani* [On the moral and civil primacy of the Italians]. The principal theses of the book were two: since time immemorial, the destiny of the Italian people was to civilize other peoples, because Rome, the centre of Christianity, was also the centre of the civilised world; by this train of reasoning,

Gioberti arrived at the proposal of a federation of all the Italian States under the presidency of the Pope. This "risky prophetism"[153] of Gioberti's opened the door to a flood of political writings; among them were *Speranze d'Italia* [Hopes of Italy], published by Count Cesare Balbo (1789-1853) in 1843 and *Degli ultimi casi di Romagna* [About the recent events in Romagna][154] by Marquis Massimo d'Azeglio (1798-1866). Balbo, Gioberti and d'Azeglio were - in that order - the heads of the first Piedmontese constitutional governments, named by the King on the basis of the Statuto granted by Carlo Alberto on 4 March 1848.

At this date, the revolution was already in full swing. On 16 June 1846, Cardinal Giovanni Mastai-Ferretti had been elected Pope, with the name Pius IX; the election of Pius IX and his first political moves seemed to confirm the liberal hopes of Gioberti's *Primato*. Starting from March 1847, a series of reforms was announced in the Papal States, in Tuscany and in Piedmont. But already in July 1847 Austrian troops occupied Ferrara, in the Papal States, after unrest caused by disappointment that Pius IX was not introducing more serious reforms. In January 1848 a revolt in Palermo freed Sicily from the corrupt rule of the Bourbons, and within a short time riots and revolts inflamed the Country; in March, after five days of street fighting (which have entered national history as the *Cinque giornate*) the Milanese patriots[155] compelled the Austrian troops to leave the town. With some reason, the King of Sardinia feared that the political movements in Piedmont that the political movements in Piedmont and Lombardy might take a republican turn; in Milan, for example, the president of the war committee was Cattaneo, a federalist and republican. With mixed feelings, [156] Carlo Alberto declared war on Austria, which by now was enfeebled by several revolts in the multi-ethnic Empire.

I have lingered over the political context of the revolutions and war of 1848-49, because those years were really crucial for the formation of a national conscience of Italian scientists, as is demonstrated by their generous participation in political and military events. Chemists and pharmacists, in particular, were active in almost all possible roles. Their revolutionary activity may be seen in many theatres of war and rebellion.

In the North, Agostino Frapolli (1824-1903) fought the military campaign of 1848 and 1849 in a volunteer corps. After the war he studied chemistry in Milan and then in Heidelberg with Bunsen;[157] in 1859 he became Head of a private school of chemistry in Milan. Tullio Brugnatelli (1825-1906), then a student, fought in the *Cinque giornate*, and in 1849, after the defeat of Carlo Alberto, participated in the desperate defence of Venice; later, in 1859, he was appointed to the Chair of general chemistry at Pavia. Antonio Stoppani (1824-1891) also fought on the Milanese barricades, at that time a young seminarist; he was later ordained and, persecuted by the Austrian authorities for his liberal ideas, went into exile in Piedmont. A capable geologist, after the annexation of Lombardy in 1859 he was appointed to a

Chair of geology at Pavia - in part a reward for his patriotism. At the beginning of the uprisings, Francesco Selmi (1817-1881), was a secondary school teacher in Modena, with a diploma in pharmacy. Deeply interested in publishing, he founded a newspaper, *Giornale di Reggio*, that supported the annexation to Piedmont of the provinces then under the rule of the Duke of Modena. After the return of the Duke he went into exile in Turin. In the Grand-Duchy of Tuscany, Gioacchino Taddei (1792-1860) was professor of pharmacology at the Hospital of Florence. He was elected a member of the revolutionary Parliament of Tuscany, and became its President; for this political role he was dismissed from all official posts after the return of the Grand-Duke. The Tuscan Provisional Government appointed Giuseppe Orosi (1816-1875) to a Chair of pharmaceutical chemistry; after the restoration, he returned to his activity of private researcher and author in the pharmacological field. In 1860, he had again a Chair, in-medical-pharmaceutical chemistry at Pisa. A victim of the return of the Bourbons to Naples was Oronzo Gabriele Costa (1787-1867), a zoologist, who lost his Chair because of his patriotic behaviour and liberal ideas.[158] In the far South of the Peninsula, in Calabria, Sebastiano de Luca (1820-1880), later friend and colleague of Cannizzaro's, participated in the revolutionary upraising, and after the defeat was sentenced to 19 years in jail, and went into exile in Paris. Finally, the young Stanislao Cannizzaro (1826-1910) also participated in every step of the revolution in Sicily. Elected member of the Sicilian Parliament, he fought as an officer of the artillery. After the revolt had been defeated he went into exile in France.

Some of these patriots, particularly adventurous, were exceptionally active; one of these was the pharmacist Dario Tassoni (1818-1885) who after having fought with Carlo Alberto's assault troops joined Garibaldi in the defence of Rome, and was promoted captain in the field.[159] Meanwhile, a very particular situation was developing in Rome. The uprising in the Eternal City struck at a scientific community that consisted almost completely of priests and members of the many religious Orders; it is not surprising that researchers as Father Francesco De Vico and the Jesuit Angelo Secchi[160] simply left town before the beginning of the long and extremely bloody siege by the French troops under Louis Bonaparte.

The story of scientists' participation in the events of 1848-49 would be seriously incomplete without a reference to the defiant attitude of the Pisan scientists. Carlo Matteucci, the physiologist and physicist, chose active politics, in Tuscany and abroad. He went as far as the Parliament of Frankfurt where, on 5 October 1848, he vehemently called for European solidarity: "*L'assemblée germanique n'a pas à soutenir en Italie des interêts différents de ceux de l'Europe*". But the appeal conflicted with the *Großdeutchland* dreams of many delegates, and the intervention of Matteucci had no concrete effect.[161] The other three components of the formidable quartet quoted above - Mossotti, the physicist, Piria, the chemist, and Pilla, the geologist - chose to take up arms. They fought against the Austrian as officers of the volunteer battalion of Pisan students; this battalion also contained the physicist Riccardo Felici (1819-1902) as lieutenant, and the young chemist Cesare Bertagnini (1827-1856) as a simple soldier; Bertagnini later became Cannizzaro's closest friend and the great hope of organic chemistry. The battalion joined the Piedmontese regular troops, whom it helped in their victory at Goito. In fact it boldly fought against overwhelming numbers of Austrian troops at Curtatone and Montanara. The struggle was very fierce near Case del Mulino; in the description of a contemporary military historian: "there was a ferocious fight, and there the illustrious Neapolitan Pilla, professor and leader (*duce*) of the students met his death". Later, in the same clash, the commander of the volunteers, Giuseppe Montanelli (1813-1862), professor of law at Pisa, was seriously wounded.[162]

This extensive participation in the fight for a more liberal regime does not mean that the scientists' adhesion to the new political ideas was unanimous; far from it. On one side, the revolutionary front was divided concerning the ultimate political fate of the Peninsula - Cannizzaro for example was in favour of the autonomy of Sicily from Naples. On the other side, some scientists were openly opposed to the new ideals of Unity. Giuseppe Bianchi (1791-1866), an astronomer, had founded the Observatory of Modena in 1827; he was secretary of the Società Italiana delle Scienze, and at the end of the revolts and of the war he condemned the struggle against the legitimate rulers with harsh words, and welcomed the victory of Austrian arms over Piedmont; on 15 August 1849, in a circular letter to all members of the Società, he stated that the Società should be "happy for the reinstatement of order over the whole Peninsula (*sopra l'intera penisola*)". On this 'happy' occasion, Italy had returned to being simply a geographical entity.[163]

Piedmont was very busy, in the ten years between the defeat of Novara (23 March 1849) and the victory of Solferino (24 June 1859). In 1852, the new King, Vittorio Emanuele II, named Count Camillo Cavour (1810-1861) President of the Council, after d'Azeglio, and the young, but politically experienced aristocrat led with great ability the intricate process that brought the Kingdom of Sardinia to take revenge against Austria. In this crucial decade, Piedmont was the only Italian State that seemed to have a chance of becoming the military, political, and cultural centre for the unification of the Country. Before concluding this aperçu of forty years of Italian history, I must add a couple of remarks about the cultural aspects of Cavour's era.

The first concerns the hearty reception that was given to thousands of refugees coming from all the other Italian States. The list of intellectuals, important and less so, who arrived in Piedmont to give their political and cultural contribution to the recovery, is long and impressive. Certainly there were several cases of intolerance against these refugees, who in many cases had been sentenced to death or to many years in jail, and thus had been compelled to go into exile. In Piedmont that intolerance came from the clerical side of public opinion, who were systematically against the Unity of the Country, which it now seemed clear could only come about through the destruction or the dramatic reduction of the temporal power of the Pope. The case of the Chairs of chemistry at Turin and Genoa was typical. Cannizzaro, after a stay in Paris in Cahours' laboratory, was appointed in 1851 as teacher of physics and chemistry at a secondary school in Alessandria. His best friend Bertagnini lived in Montignoso, and was Piria's assistant, in Pisa. The three chemists collaborated actively, amid great difficulty, including the fact that Alessandria was in Piedmont, Montignoso in the Duchy of Modena, and Pisa in the Grand- Duchy of Tuscany. Their fame grew and, in 1855, by a political manoeuvre, Piria was appointed to the Chair of chemistry at Turin, Cannizzaro obtained the Chair of chemistry at Genoa and, after a short time, Bertagnini inherited his master's Chair in Pisa. The manoeuvre did not pass unnoticed, and Piria and Cannizzaro immediately came under attack in the clerical press for their revolutionary pasts, and for their anti-clerical stance. These attacks urged Cannizzaro to publish, and thus they were an extrinsic cause of the publication of the now famous *Sunto di un corso di filosofia chimica*.[164]

One last point I want to make concerning science in Piedmont is to mention some exponents of the continuing scientific tradition within the various Army corps. Giovanni Cavalli (1808-1879) studied in the Military Academy of Turin, was a pupil of Plana's and became a teacher of artillery and machine design in the same Academy. His most famous innovation, which had a European resonance, was the adoption of rifled barrels in artillery pieces.[165] I have already mentioned the general Luigi Federico Menabrea; he was professor of building science at Turin University, where he taught a generation of specialists, including those who carried out one of the most important engineering projects of the time, launched in 1849: the construction of the 12 kilometres long Frejus tunnel between Piedmont and France. The feasibility of the project depended on new drilling machines powered by compressed air designed by Germano Sommelier (1815-1871). Finally, Count Paolo Ballada di Saint-Robert (1815-1888) must be mentioned in connection with the problems of thermodynamics posed by Sommelier's compressors. By profession, until the age of 42 years, Saint-Robert was an officer in the Sardinian Army; he taught at the School of Artillery in Turin, and became an internationally known expert in ballistics. His best contributions in thermodynamics appear in his *Principes de Thermodynamique*, where he discusses at length the problems of the expansion of gases under the conditions found in fire arms.[166] On more general ground Saint-Robert's work has been placed within the military-mechanics tradition of Sadi Carnot, whose work he was an acute critic.[167]

After 13 years of work, the Frejus tunnel was opened to railroad traffic in 1871.[168] By then Italy was unified, and scientists such as the chemist Cannizzaro and the physicist Pietro Blaserna (1836-1918) had been appointed to Chairs at the reformed University of Rome. Cannizzaro was invited by Edward Frankland, as President of the Chemical Society, to give the Faraday Lecture in 1872; the Italian chemist accepted the invitation, and wrote to his English colleague about the establishment of a new large chemical laboratory in Rome. I have used Frankland's enthusiastic answer as the epigraph for this section. For the British observer, the new laboratory was "a glorious idea": the scientific *Risorgimento* had in fact begun in Rome.

2. LOOKING FOR A CENTRE

In this part of the investigation, rather than following the chronological order, as I did in the preceding sections on the relationships between science, culture and political power, I have chosen a different narrative strategy, in order to emphasise the multi-linear nature of real historical processes, and the continuous mixing of innovations, traditions, and transitions. Of the following subsections, the first two look at two great innovations in the history of Italian science: the establishment of a 'national' academy and the inauguration of 'national' meetings of scientists. The third is primarily about the substantial tradition in astronomical research, and the fourth and last is on the difficult problems of scientific journals, and the transition from eclectic local journals to 'national', specialised ones.

2.1 Lorgna's Società Italiana delle Scienze

At the end of the XVIII century, scientific and literary culture in Italy inhabited hundreds of academies, spread over the whole Peninsula. It is significant that the initiative of a scientific society with 'national' scope was born in a 'provincial' town of the Republic of Venice. This new and unusual society was the the ambitious conception of a mathematician of the Veneto, Anton-Mario Lorgna (1730-1796). By profession he was an officer in the Corps of Engineers, and taught mathematics at the Military School of Verona, his birthplace, where the Società Italiana delle Scienze grow up.[169] Lorgna shared responsibility for to the actual foundation of the Società Italiana with four friends: Barletti, Boscovich, Malfatti and Spallanzani.[170] I will here only recall that Spallanzani was at Pavia, and that Boscovich had been working in Paris since 1774; he was to return to Italy, to Bassano, in 1782, after the circular letter of 1781, which in practice established the first core of the Società Italiana. A few words on Barletti and Malfatti may be useful.

Carlo Barletti, ordained as a *scolopio* (i.e. in a particular Roman Catholic Order), was a professor of physics at Pavia University, last mentioned when he was jailed in 1799, when the Austrians returned for a time to Lombardy. On 23 April 1784, when the Society was well launched, he in a letter to Lorgna he declared himself confident of "the Patriotic attitude of the true Italian Authors", and Lorgna, the founder, as "the foundation, the key-stone, the centre of the splendour of Italy".[171] Within the flowery eighteenth-century style, the quest for a national centre, at least for the scientific community, is very clear. A month later, on 25 May 1784, Barletti wrote emphasizing the necessity for strict control of the quality of contributions to be published by Lorgna's Società Italiana. In the letter Barletti also expressed concern about certain unjustified "aura of celebrity", and about the "clamour that is created by the public newspapers"; these criteria would obviously be prejudicial for the selection of members. The physicist added another important remark: "We judge the merits of people from the whole body of knowledge we have about them, but the Foreigners will not do so, and they will judge the Academicians and the Italian Academy from the memoirs published by each member in the Proceedings".[172] Thus, for Barletti, the aim of reaching an audience abroad and gaining an international reputation was prominent.

The mathematician Gianfrancesco Malfatti (1731- 1807) was librarian to a nobleman of Ferrara and, after the reform of 1771, professor at the University of that town, a possession of the State of the Church in North Italy.[173] thus he was probably among the first correspondents of Lorgna to be informed about the project for the new Academy. On 28 June 1766, Lorgna wrote him that in Italy a periodical newspaper was necessary "like that published in Leipzig".[174] In 1776, the project for a national Society was mature, and Malfatti wrote to Lorgna that "the idea of forming an Academy for all the Italian *Letterati* [was] noble and glorious"; but he was seriously concerned about the real value of the scientific contributions that might reach the Society from mathematicians, anatomists, and - in particular - naturalists.[175] In this sense, Malfatti's concern was similar to that expressed by Barletti, whose sincere fear was that: "We must be careful that we don't get sent *warmed-over hash (cavoli riscaldati)*".[176] Lorgna's national initiative was not the only one in which Malfatti participated.[177]

As I have said, Lorgna send the first circular letter on 1st March 1781; it clearly concentrates on the need of publishing quickly (avoiding the delays of publishing abroad) and in a handsomely printed format. He promised to take care of both these requirements at his own expense, and toward the end of the letter he looked forward to the creation of a "Body united only by the cement of love for our Country (*amor patrio*) and of free natural genius".[178] The answers, kept on file by Lorgna, included positive comments from many of the addressees.[179] But not all Lorgna's proposals gained the immediate agreement of the members-to-be, in particular the use of Italian in the publications of the new Society. Although this proposal was only implicit, in the proofs of a possible frontispiece for the *Memorie*, sent with the circular letter, it provoked reactions from several colleagues and friends. Later, in 1782, Boscovich reacted to the proposal to translate a paper of his into Italian, before publication in the new-born *Memorie di Matematica e Fisica della Società Italiana*; in particular he felt, as he said in a letter to Lorgna, that "very few abroad will read the Dissertations" if they are published in Italian. In another letter to Angelo de Cesaris (1749-1832), an astronomer at the Brera Observatory, who also resisted the use of Italian, Boscovich wrote that the translation into Italian would restrict "the use to Italy alone. The Italian of the mathematicians is not known [in Paris] nor in England, nor in Germany".[180] In the same period, a Piedmontese chemist, Count Carlo Lodovico Morozzo (1743-1804), wrote candidly to Lorgna that he knew French better than Italian, and thus feared not to be able to write his papers in this language. On 6 June 1782, Morozzo was reassured by Lorgna ("be of a calm mind"[181]) and his contribution was published in the first tome of *Memorie*.[182] Morozzo did join the Società in 1782, a year after the 'first wave', which included the following *letterati*: G.Malfatti, C.Barletti, T.Bonati, L.Lagrange, F.Fontana, G.Fontana, M.Landriani, G.Riccati, G.A.Saluzzo, L.Spallanzani, A.Volta, G.V.Zeviani, G.Torelli and L.Ximenes. These names are from the *Catalogo* of the first 40 national members of the Società Italiana, which was published in 1786, after it had been agreed that the number of members should be limited to forty; in the list, the surnames of the members were ordered by the year of their admission to the Società. The names of Torelli and Ximenes did not appear because they had died before publication of the *Catalogo*.

This list[183] merits our attention from several points of view. Given the national ambitions of Lorgna's 'brain-child' it is of interest to look at the local origins of the members. But it has to be borne in mind that Region or State do not coincide. Penso, for example, 'assigns' six scientists to Emilia (a Region in the north of Italy). This is not the most revealing classification, however, for there was no such thing as Emilia. Today's Emilia was at that time divided into three different states: following the old Roman *via Emilia*, a traveller, from Milan to Rome, would have needed passports at the frontiers of the Duchy of Parma, the Duchy of Modena, and Papal Bologna. Penso's six *letterati* lived in fact under four very different governments. G.B.Venturi (1746-11822), who was to be politically very active in the Napoleonic period, was teaching Physics in the Duchy of Modena.[184] Three lived in the Papal States: T.Bonati, physician and mathematician, Sebastiano Canterzani (1734-1819), and the old astronomer Eustachio Zanotti (1709-1782), retired from the Chair of astronomy at Bologna in 1760.[185] L.Caldani (1725-1813), the successor of the great Morgagni, held the Chair of anatomy in Padua,[186] in the *Serenissima* Republic of Venice; and, lastly, L.Spallanzani was a professor at the University of Pavia, just south of Milan, his salary paid by the Imperial and Royal Government of Austria.

In terms of scientific life and culture, the States are the relevant division. Overall, in Northern Italythe Society counted a total of 30 members: five members in the Kingdom of Sardinia (all of them in Turin); twelve in Lombardy under Austrian rule; eight in the territories of Venice; one each in the Duchy of Parma and the Duchy of Modena, and the three mentioned above under Papal government. In Central Italy, that is to say in the Grand-Duchy of Tuscany, there were seven members. The remaining three members were all in one city, Naples, capital of the Kingdom of the Two Sicilies. Apparently, after the Grand-Duchy of Tuscany, scientific culture had only one strong-hold, Naples, a town that had been a great cultural centre for uncountable centuries.[187] Lorgna, in his correspondence, makes the most of this span - passing over Rome, which conspicously sent no members; he writes (to Spallanzani) of "our Country from Naples to Turin", and to Canterzani of "members of Italy, from Naples to Piedmont").[188]

We may now try to put Lorgna's Società in the European context, looking for confirmation of the international value of the men chosen by Lorgna and his friends. Several years ago Marie Boas Hall published an analysis of the Italian membership of the Royal Society,[189] in the period 1667- 1797. Comparing the Italian members of the Royal Society with the members of the Società Italiana in the year of publication of the *Catalogo* and in the following year, some interesting results emerge (see the following Table). The first is clear: the Italian *letterati* gathered by Lorgna was a group of unquestionable excellence in the European context, at least as the Royal Society assessed scientific excellence at the time. The second is that the Società Italiana promoted the international visibility of its members, because of the six Italian men of science elected to the Royal Society, after 1781 (Lorgna's Society foundation) until the end of the century, five were members of the Società.

It must be added that London was not the European scientific capital with which the Members of the Società were most familiar. This was undoubtedly Paris, where Cagnoli presented the first volume of the *Memorie* to the *Académie*, and reported to Lorgna the positive reactions of the Parisian scientific community.[190] In the following years, the Società went through dynamic and slack moments, depending on the intentions, capability and political attitude of the President in charge. It maintained a high level of membership, published interesting papers but was by no means an active centre of organization or of political presence for the scientific community.

Lorgna's Società Italiana: a view from London		
	Date of membership	
	Royal Society	Società Italiana
B.Oriani	1795	1785
G.Fontana	1795	1781
A.Volta	1791	1781
A.Scarpa	1791	1783
A.M.Lorgna	1788	founder
L.Caldani	1772	1785
L.Spallanzani	1768	1781
G.F.Cigna	1764	1784
S.Staticò	1764	1782
G.Saluzzo	1760	1781
E.Zanotti†	1740	1782

2.2 The Meetings of Italian scientists

> I Congressi degli scienziati italiani [...] ebbero una doppia efficacia: 1) riunire gli intellettuali del grado più elevato, concentrandoli e moltiplicando il loro influsso; 2) ottenere un più rapida concentrazione e un più deciso orientamento negli intellettuali dei gradi inferiori.
>
> *Antonio Gramsci*, Il Risorgimento[191]

Between 1839 and 1847, nine scientific congresses were held in nine different Italian towns, including the capitals of the Kingdom of Sardinia (1840), the Grand-Duchy of Tuscany (1841), the Italian provinces in the Austrian Empire (Milan in 1844, and Venice in 1847), and the Kingdom of the Two Sicilies (1845). Beyond any doubt, this series of meetings was the most important public initiative of Italian scientists during the Risorgimento, and probably during the whole (social) history of Italian science. Historians' attention has been attracted by these *Riunioni*, and they have considered them especially from the political point of view.[192] Here we are also interested in these meetings as they appeared to many contemporaries, as a way of celebrating and organizing the scientific community. The meetings of naturalists and physicians (or of scientists, for brevity) had begun in Switzerland in 1815, when 36 people met in Mornex. This early start is not surprising, if we remember that the multi-lingual Switzerland developed a 'nationalistic' mind-set at least as early as the first decades of the XVIII century; in 1727, in Zurich, Johan Jacob Bodmer established the first historical society dedicated to the history of a Country (Helvetische Gesellschaft).[193] A few years after the Swiss meetings began, in 1821, the romantic and liberal biologist Lorenz Oken participated in one of them, and proposed a similar initiative to naturalists and physicians living in the multitude of German States. The figure of the founder aroused much suspicion, but nevertheless the Gesellschaft Deutscher Naturforscher und Aertze was established for the purpose of reciprocal "acquaintance between members";[194] in 1822, only 15 people met for the first congress in Leipzig, but within a short time - as we will see - the authorities were able to control all possible political aspects of the meetings, and the audience grew. As is well known, in 1830 the mathematician Charles Babbage published his *Reflections on the decline of science in England*; in the following year he and other 'gentlemen of science' established the British Association for the Advancement of Science, and held their first meeting in Oxford.[195] Three years after the July Revolution, in 1833, the French scientists held the first of their series of meetings, with 220 participants, in the provincial town of Caën, where A. De Caumont, the promoter of the congress, lived.[196] And at last, a generation after the Swiss Confederation, in 1839 Italy also held her first congress of scientists, in Pisa, with 421 participants. (Interestingly, in 1840 the Hungarian scientists also held their first meeting.[197]) Before considering the Italian *Riunioni* more closely, we may look at this process by which scientific communities organized themselves, so as to get a clearer picture of the international context in which the Italian initiative took place.

It is obvious at first sight that the process was European[198], and that some of the participants themselves already saw it in this sense. The second is that 'nations' that were very different in their political and geographical situations were involved in the different steps of the process. The borders of Italy and Germany were not defined at all, and France did not then include Savoy. Considering the complete list - Switzerland, Germany, Great Britain, France, Italy, Hungary - in four cases the name of the nation corresponds to an existing political and cultural reality, and in this sense Switzerland and France are unproblematic. In the case of Great Britain it has been remarked that Babbage, the 'declinist' *par excellence*, held up France as the model for Britain to follow, ignoring the situation in the Scottish Universities, and the richness of English provincial science.[199] For Babbage the 'national centres' were the old traditional ones in the South of England. Hungary, too, was a political entity, but only as a separate Kingdom under the (imperial) crown of Austria, and this is probably the country where the problem of 'nation' was most present. Germany and Italy are different from the others, and similar to one another, because these 'nations' were divided into many independent states and the basis for the meeting was essentially cultural and linguistic. The political, liberal intentions of the promoters were in some cases manifest, but the participants were of every possible political shade. Two or three further points are relevant to this research about the dialectics between centre and periphery. First, in all the six cases here considered, without exceptions, the meetings were itinerant, and if we plot the national meetings on the map of Europe, on the eve of the 1848 blow-up, we find thousands of 'scientists' travelling over a large part of the Continent, in order to meet their peers. This type of migration seems to point to a sociological interpretation much more than to a political explanation; that is, European scientists in the first decades of the XIX century were actively constructing their social image, role and function as scientists. Above all, their collective movements demonstrate that for them also, and at that time also, the problem of prominent scientific centres was a real one.

The first Italian meeting was promoted and organised by a small group of six scientists; one of them, a man with the remarkable name of Bonaparte, had participated in the Freiburg meeting of the Gesellschaft Deutscher Naturforscher und Aertze, and - exactly as Oken had done - on hia return to Florence began to mobilize the notable scientific resources of the Grand-Duchy, in order to organize an Italian Society on the model of the German one. The aim of the new organisation was to "support the progress and diffusion of sciences and their useful applications"; among the sciences there was explicit mention of "Medicine and Agriculture, so useful to humanity". The promoters, in the order of their signatures to the circular letter that announced the *Riunione,* were: Prince Carlo L. Bonaparte, politician and nephew of Napoleon, present as a zoologist; Vincenzo Antinori, Director of the "I.e R." (Imperial and Royal) Museo di Fisica e Storia Naturale of Florence; G.B.Amici, here as Astronomer to His "I.e R.". Highness the Grand-Duke of Tuscany; Gaetano Giorgini, engineer and professor at Pisa University, as *Procuratore Generale* of the "I.e R." University; Paolo Salvi, professor of Natural History at the same University; and lastly Maurizio Bufalini, professor of Clinical Medicine at the "I.e R." Arcispedale of Florence. With the evident exception of Carlo Bonaparte, all the scientists worked in Imperial Royal institutions, announced by the tediously repeated abbreviation "I.e R.". This display of official title might have reassured many of the addressees of the letter,[200] but not always their political rulers, in capitals as different as Turin, Vienna and Rome. In Turin, it was suspected that the whole affair might be a manoeuvre of Grand-Duke Leopoldo II to gain points in the contest for supremacy in Italy, so the Piedmontese Government manoevred in order

*The first eight Riunioni of Italian scientists**

Date	Seat	Participants	Date	Seat	Participants
1839	Pisa	421	1843	Lucca	446
1840	Turin	573	1844	Milan	1159
1841	Florence	900	1845	Naples	1545
1842	Padua	514	1846	Genoa	1062

* Source: B.Bertini, *Relazione*

to have Turin as the seat of the second meeting. Probably this move had a positive effect on the Vienna Chancellery, which overcame its strong suspicions about the liberal and patriotic significance of the initiative, and permitted the scientists resident in the Italian provinces to go to Pisa. The Holy See remained unmovable, however, and scientists living in the States of the Church were forbidden even to correspond with the organizers.[201]

It is superfluous here to follow all the nine *Riunioni*, but some of their general features may be outlined here. Only eleven people were present at the first five *Riunioni*; they included Carlo Bonaparte, the Director of the Civic Museum of Milan, Gianalessandro Majocchi, who was interested in scientific publishing (see below), several aristocrats who were also great land-owners, a few naturalists and one or two quasi-lunatics. It is evident that there was no core of scientists of the physical sciences who would, or could, travel every year to spend two weeks at a congress. Ignazio Cantù, a patient man, counted all the various participants in the first five meetings, and obtained 2,216 names; 228 participants were foreign, 685 were resident in the Grand-Duchy of Tuscany, 565 in the Sardinian States, 460 in the Austrian Monarchy, 140 in the Duchy of Lucca, 45 in the States of the Church, 25 in the Kingdom of the Two Sicilies, 25 in the Duchy of Modena, 25 in the Duchy of Parma; 15 were Italians resident abroad. No meeting had been held in the Austrian Monarchy (to use Cantù's denomination) so the large number of participants from Lombardy and the Veneto is significant. As shown in the Table below, the audience at these meetings was large, even by the present-day standards. The problems that the large participation posed to the so-called authorities will be described shortly, but first I will consider the data published by the Dean of the Turin Medical Faculty, Bernardino Bertini (1786-1857). He participated in many *Riunioni* of Italian scientists and, with official roles in the medical section, in the analogous meetings of the French scientists. He took a direct part in the efforts of European scientists to became more visible to their Governments and to public opinion, and after 1848 he was to become a member of the Piedmontese Parliament (conceded by King Carlo Alberto), so it is clear that he was also interested in the political aspects of the Italian scientific community. In the first months of 1847, he published a report on the XIV meeting of the French society, that was printed in the *Giornale delle Scienze Mediche*, the official organ of the Reale Accademia Medico-Chirurgica of Turin; the paper circulated also as a separate reprint.[202] In this formal context, he defended the *Riunioni* from press attacks;[203] adversaries had suggested that the Italian meetings "gave rise to sarcasm, became matters of derision and scorn (*irrisioni e vituperii*)". In perfect *Risorgimento* style, Bertini responds that just a few weeks before, in Genoa, the *Riunione* had been christened: *Parlamento della Sapienza Italiana* (Parliament of Italian Wisdom), and adds quantitative data on the Meetings of Scientists held in the various Countries: in Germany the biggest audience had been in Brunswick (652 participants, 1841); in the France the maximum was in Lyon (990 participants, 1841) and Strasbourg (1100 participants, 1842).[204]

Because Bertini, as well as other colleagues of his, was interested in the French meetings, the reports of the medical sections arrived regularly in Turin. One of these reports is particularly interesting in the context of this research, both for its content and because it was bound in a volume with other similar reports, from meetings in Italy and abroad. The report in question is on the medical section of the Strasbourg meeting, the section in which the number of participants reached the maximum of participants. In the introduction before the technical report, the author (the French physician Pétrequin) affirms that the congress has been an enormous success because Strasbourg is a *pont jeté entre la France et l'Allemagne*; the congresses have become acclimatised in Germany and in Switzerland, so they have acquired an European character. He concludes by returning to the exceptional audiences of the meetings of Strasbourg, at the border with Germany, and of Lyon, near the border with the Kingdom of Sardinia, and states that those "border-towns (*villes-frontières*) call for the scientific fusion of the European nationalities".[205] These considerations were expressed and published 'in real time'; the second Italian meeting was held in the days 15-30 September 1840 in Turin, and within few weeks Bertini sent his colleagues reprints of his report, published in the medical journal *Effemeridi delle scienze mediche*, November 1840 issue. The Turin meeting, to which I will return later, saw the presence of several Swiss scientists, and in the last months of 1840 the physicist August De La Rive, of Geneva, published the first part of a detailed report of the meeting in the very widely circulated *Bibliothèque universelle de Genéve*, and the second part in 1841.[206] It is to be remembered that the Lyon meeting of French scientists was held a year later than that in Turin; Oken was present at the first meeting in Pisa, Babbage and De Caumont participated in the Turin meeting. Thus the founders of the three most important scientific associations in Europe were involved in the birth of the Italian movement; finally, as I said above, Cantù remarked on the presence of 10% of 'foreigners' at the first five meetings. This interplay of presences and references means, once again, that the *Riunioni* were an integral part of a cultural movement that looked at Italian problems (scientific and non-scientific) in a European context.

Of course, Italian scientists considered the economic and social problems of Italy (or better of the different Italian States) through the eyes of their own social status. This status was analysed some years ago, and the sociological results confirm the impression gained from the biographies of community leaders in the different fields. In synthesis, three out of four participants were bourgeois, and the remaining quarter were aristocrats; for obvious political reasons, only a meagre 2% were clergymen, far below the cultural presence of the clergy in Italy. Half the participants came from Universities and Academies, and 90% of them had University degrees.[207] At the same time, however, almost all political power and much of the cultural hegemony was by then in the hands of the aristocracy. That hegemony can be illustrated by two distinguished authors: the most famous Italian novel of the XIX century (*I promessi sposi* [The Betrothed], first edition, 1827), was written by Count Alessandro Manzoni (1785-1873), whose mother was the restless daughter of Marquis Cesare Beccaria, probably the most famous Italian intellectual of the XVII century; and the greatest Italian poet of the XIX century was Count Giacomo Leopardi (1798-1837), whose *Operette morali* appeared in 1827, and his *Canti* in 1831.[208] The two Romantic writers lived in very different milieus; Manzoni in cosmopolitan, modern Lombardy, Leopardi in an insular district of the States of the Church, but both had the same "poetic greatness, which is measured by the scope and intensity of their cultural perspective, by the perspicuity of their view of the problems, feelings and ideologies of the contemporary world".[209] This literary diversion has brought us nearer to an important aspect of Italian culture: it sustained itself in a multitude of small towns, throughout Italy, in Piedmont as well as in the States of the Church and in Sicily. The distance of most of this culture from science was not geographic, but ideological; a sort of epistemological proscription denied that scientific knowledge could have any deep value.[210]

We turn next to the scientific structure and content of the meetings. From the first meeting in Pisa until the fourth in Padua, the meetings were organised in the following sections: 1) Agronomy and Technology; 2) Zoology, Comparative Anatomy and Physiology; 3) Physics, Chemistry and Mathematics; 4) Mineralogy, Geology and Geography; 5) Botany and Vegetable Physiology; and 6) Medicine with a subsection for Surgery. The list merits a few comments. The coupling of agricultural and technological problems, in that order, clearly means that, in the Italian - or the Tuscan - context, engineering was seen in relation to the development of the territory. This priority of agricultural interests was clearly reflected in the Italian market of scientific books: in the years around 1840 the distribution of topics was such that more than 40% of books were on agriculture, 16% on physics and only 8% on chemistry.[211] Comparative anatomy is an interesting title: this subject was somewhat neglected in Italy, being considered as leading to transformist ideas, not at all welcome in the States restored by the Vienna Congress. Franco Andrea Bonelli (1784-1830), for example, as professor of zoology in Turin, published and lectured on an uncritically traditional zoology, and reserved hir thoughts on 'philosophical zoology' for the privacy of the laboratory and personal correspondence.[212] The section on physics, chemistry and mathematics split in 1842, when chemistry was given its own section. The following year, in Lucca, a similar division was proposed between physics and mathematics. Majocchi, for example, spoke in favour of the separation, recalling the limited mathematical capability of "distinguished physicists (*esimj fisici*)" such as Volta, Davy and Faraday, but, due to the understandable opposition of the astronomers, the proposal was rejected.[213]

Massimo Galluzzi has analysed the content of the contributions in the physics and mathematics section, and has concluded that the meetings were serious occasions for contact, and in several cases for confrontation, between the diverse more or less 'modern' currents flowing through the rich terrain of Italian mathematics.[214] The chemistry contributions to the first eight meetings have also been analysed. A direct comparison with the scientific papers published in a leading journal like *Annales de Chimie et Physique* may give the impression that many of these communications seem outdated; they should probably be seen as reports about chemical investigations motivated by practical interests, such as pharmacy, medicine or agriculture. At the theoretical level, the Italian chemists were perfectly well informed about the most recent developments, but the consequences on the style of research were negligible. The frequent, and sometimes lively, discussions about the atomic theory and the constitution of bodies were in fact more pertinent to teaching than to research.[215]

Cantù estimated that about 10% of the participants were foreign, and we have already seen the significance of the presence of Oken, De Caumont and Babbage. At the Turin meeting, Babbage presented his calculating or analytical machine, and several *memorabilia* of the presentation are kept in the Academy's archives.[216] Other participants contributing to the web of relationships between Italian scientists and their European colleagues were the astronomer and statistician Adolf Quetelet from the Observatory of Brussels; the great Swiss botanist Augustin-Pyramus De Candolle, from Geneva; the chemist Christian Schönbein, from Basle; the mathematician Karl Gustav Jacobi, from Berlin; Friedrich Tiedeman, editor of the *Zeitschrift für Physiologie* and Director of the Institute of Anatomy in Heidelberg.[217] In short, the *Riunioni* were both a mirror of the Italian scientific community and a bridge towards European science: the obscure isolation of the Restoration years had ended.

2.3 Traditions: Astronomy and Other Sciences

In a sense it is obvious that, in any account of Italian scientific traditions, the most immediate references are to mathematics, astronomy and medicine. All three disciplines have been pillars of European scientific culture, and in any epoch have seen intense exchanges of men, instruments and ideas between the European countries. I have chosen here to follow the development of astronomy, essentially because of its rich connections with several aspects of the international scene, including the market in scientific instruments. Four astronomers, belonging to four successive generations, allow us to follow the development of the Italian scientific milieu from the end of the XVII century up to Unification.[218]

Giuseppe Piazzi (1746-1826) was born in a small town, Ponte di Valtellina, then in the Swiss Confederation; son of a well-to-do family, he was sent to study to Milan, where he was ordained *teatino* (a priest of one of the many Roman Catholic Orders); afterwards he studied in Turin and Rome, obtained a degree in mathematics and philosophy, and in the decade 1769-79 he taught or preached in several towns, going as far South as Malta.[219] An important decision taken by Bourbon rulers of the Kingdom of the Two Sicilies changed his way of life: in 1779 they established the University of Palermo, and in 1780 Piazzi was called by the Viceroy to teach 'sublime calculus'. Then, in 1786 the Government decided to set up an Observatory in Palermo, charging Piazzi with its planning and realization. He decided to visit the places where the best astronomical instruments were built, and where the best astronomers were probably working. First in 1787 he went to Paris, where he was the guest of one of the best astronomers of the time, De Lalande, and met Delambre, Pingré, and Méchain. The following year, in England, he came to know Maskelyne and Herschel, and spent some time at Halifax with Jesse Ramsden, a genius of precision mechanics and a vigorous entrepreneur.[220] Piazzi ordered several instruments from Ramsden, the most important being an altazimuth telescope with a circle five foot in diameter.[221] In the new observatory at Palermo, built in a tower of the Royal Castle, Piazzi began a long series of observations on star positions, to catalogue them as precisely as possible, and it was in the course of this project that he made his most important discovery. During the night of January 1st 1801, while measuring the data of a certain star catalogued by La Caille, he noted that that star was preceded by another one of the eighth magnitude, and the following night observed that this latter 'star' had changed its position. He concluded that it was a planet, and he followed its orbit before communicating his discovery to Barnaba Oriani (1752-1832), the Director of the Observatory of Brera, and to Johann Elert Bode, the Director of the Berlin Observatory. Our *teatino*, passing his nights in the southern-most Observatory of Europe, made his discovery within in a specific European context. The discovery of Uranus by William Herschel in 1781 had convinced Bode that his 'law' relating the position of the planets with the terms of a geometrical series was right: it followed that a planet with an orbit between the Earth and Mars had yet to be discovered. He and Franz Zach, a Hungarian astronomer, called for a meeting of astronomers in Lilienthal, where a systematic search was organized, divided among 24 observers. Piazzi was one of the 24, but he had in fact made the discovery of the small new planet before being told of the joint project. When the news of his discovery reached Bode the new object was disappearing in the sunlight, and the positions observed seemed too few to calculate the orbit. At this point, a young mathematician, Karl Friedrich Gauss, then 24 years old intervened. He was able to calculate the orbit using the method of least squares - his own invention - and indicated the probable present position of the new object. During the night of 31 December Zach observed the 'planetoid' in that position, triumphally closing a year in which, all over Europe, the community of astronomers had cooperated to remakably good effect.[222] Piazzi named the new 'planetoid' (or 'asteroid' as Herschel preferred to say) Ceres. His fame is principally due to this discovery; however, he continued his accurate observations of star positions, and published a catalogue of 6,748 stars in 1803, which enjoyed a great success; and his international stature was confirmed when more than 5,500 of his observations were edited by Bode and published in Berlin.[223]

The chain of events that brought Giovanni Plana (1781-1864) to astronomy and geodesy was very different from Piazzi's, as may be imagined from his date of birth, just a few years before the French Revolution. As a teenager, when Piedmont was invaded by Napoleon in 1796 he planted an *arbre de la liberté* in the courtyard of his school, in Voghera, and was compelled to take refuge in Savoy. In Grenoble he studied in the new École Centrale, where he became a friend of H.Beyle (the future Stendhal). In 1800, he was a French citizen, and won an open competition to enter the École Polytechnique, where Lagrange, Laplace, Monge and Legendre were among his teachers. Three years later, he returned to Piedmont, where he taught mathematics at the School of Artillery in Turin. In 1811, recommended by Lagrange, he became professor of astronomy at the University of Turin.[224] He had developed an interest in theoretical astronomy some time earlier, during contacts with the astronomers at Brera, then still led by Barnaba Oriani. In that stimulating milieu he became a friend of Francesco Carlini (1783-1862), future co-worker in his most important scientific enterprise. Carlini's life had been less adventurous than Plana's: he had grown up in Palazzo Brera, where his father had served as a librarian and curator of the great collections that were the nucleus of one of the most important Italian public libraries, the Braidense.[225] Stimulated by Oriani, the two young astronomers engaged on a very ambitious project: a complete mathematical theory of the highly perturbed motion of the moon. Oriani's suggestion, probably made in about 1813,[226] was really up to date: in 1820, Pierre-Simon de Laplace would propose just that problem to the Académie des Sciences as the theme for the astronomy prize. Plana and Carlini presented their results, and won the prize jointly with a French researcher, Damoiseau; this victory was a great success, reaching into the very inner sanctum of the French *patronage* system. The theory of Plana and Carlini was never published in that form; the two astronomers continued to work on the problem, but their collaboration broke down, and Plana started the work again alone: in 1832 he published the conclusions of his research in three large volumes.[227] The intrusion of foreigners on the sacred soil of the *mécanique céleste* did not remained unnoticed, and Plana found himself engaged in a long polemic with Laplace and his school (Poisson in particular), initially together with Carlini and in a second time alone. The polemic reached its climax with a long communication by Plana read to the Astronomical Society of London on 9 December 1825.[228] The Astronomical Society rewarded Plana with a gold medal after the publication of the *Théorie du mouvement de la lune* in 1832.[229]

Before their collaboration ended, Plana and Carlini also worked together in another endeavour, not only theoretical. Their respective Governments adhered to the French proposal of continuing the triangulation along the 45° parallel, which by 1821 had reached the western border of Savoy. The chain of triangulation was almost complete, from the Gironde to the Adriatic sea, with the notable exception of the most mountainous districts between Turin and the Savoy border.[230] The Government of the Kingdom of Sardinia entrusted the Piedmontese Army topographers with the task of completing the triangulation, while the Austrian Government placed Carlini alongside Plana for the geodetic and astronomical measurements. In only three years, the Piedmontese Army completed the topographic campaign; the results were published in two large volumes in 1825 and 1827, the second one compiled by the two astronomers alone.[231] In the best astronomical tradition, Carlini and Plana were also active in related fields. Carlini was interested in meteorology, a field in which he was able to correlate meteorological observations with the rotation of the sun;[232] Plana was also a good mathematician, and made contributions to infinitesimal analysis and mathematical physics.[233] Plana, it is to be remembered, was Cauchy's principal sparring-partner during the French 'rigourist''s stay in Turin. In 1830 Cauchy published a long article in the *Biblioteca italiana*, a sort of philosophical manifesto against mathematics *à la* Lagrange, and in many conversations with Plana he reasserted that the methods used by astronomers were without any rigour; in October 1831 he presented a long *Mémoire sur la mécanique céleste et sur un nouveau calcul appelé calcul des limites* to the Turin Academy, with many interruptions from Plana; the *Mémoire* was never published by the Academy, but circulated only *litographié*.[234] Plana's defence of Laplacian mathematical physics demonstrates that his polemic with the French researchers of the École Polytechnique was a polemic within one and the same school.

Another Italian scenario was that in which Angelo Secchi (1818-1878) lived; he was one of the most interesting protagonists of scientific life in Italy in the mid-XIX century. The scenario was that of the Restoration, and of its 'evolution' in the States of the Church under the pressure of Risorgimento events. When Pope Pius VII returned to Rome in May 1814, he immediately restored the notorious 'trinity' of instruments for cultural control: the Inquisition, the *Index librorum prohibitorum*, and the powerful Society of Jesus. Reggio Emilia, Secchi's birth place, again came under the rule of the Holy See, and the local Jesuits' College was reopened. Angelo Secchi studied in this college, and when he was fifteen he joined the Society of Jesus, a step that obviously also deeply conditioned his scientific life. He began a career, normal for a brilliant Jesuit, as teacher (of grammar and philology, and later of physics and mathematics). He was in Rome when the Revolution reached that city, in March 1848, and felt compelled to flee to England and, later, to the United States, where at Georgetown University, he was able to begin astronomical research. When the Revolution in Rome was bloodily put down by the French troops, Secchi was called to take up the post of director of the Observatory of the Collegio Romano. On his way back to Italy, Secchi stopped in Greenwich and Paris; in Rome he was able to build a new observatory and to furnish it with important new instruments.[235] Before looking at his scientific and philosophical activity, a last biographical particular requires mention. In 1870, during the Prussian-French war, Italian troops stormed the walled part of Rome, and after more than a millennium installed a secular Government in the city. A group of scientists interested in establishing a first-rate University in Rome caused the Italian Government to offer a Chair of Astronomy to Secchi, but the Jesuit astronomer was compelled by his Order to refuse; he was able to maintain the Directorship of the Observatory of the Roman College, although the building was confiscated by the Italian State.[236]

As Redondi aptly remarks, "In Astronomy Secchi developed the best outcomes of the profound experimental vocation that characterizes Italian scientific culture in the first half of the XIX century".[237] Secchi was truly a man of his times from many points of view: for example he built, and presented at the Paris Exhibition of 1867, a fully automated recording meteorological station, which could measure air-pressure, temperature, wind speed and rainfall. The elaborate apparatus, of Victorian design, was admired by Napoleon III, and Secchi was accorded the *Légion d'honneur*.[238] But the field in which the Jesuit astronomer made his greatest contribution was the foundation of astrophysics. Effective spectroscopic study of the stars became possible after the turning-point in knowledge of atomic spectra achieved by the formidable couple of Kirchhoff and Bunsen. In 1859, just the the same year in which the two German scientists made their discovery, Giovanni Battista Donati

(1826-1873) had become Director of the Observatory of Arcetri, near Florence, succeeding Amici, the great optician. Donati was immediately interested in the application to astronomy of the spectroscopic recognition of the chemical elements, and about 1860 tried to observe star spectra, helped in the necessary instrumental innovation by Amici. The first results for the spectra of 15 stars were published in 1862 by Donati, but the experimental apparatus was unsatisfactory and Amici invented a particular compound prism that permitted the observation of the spectra without the usual deviation of the axis of propagation of the light.

Amici's system was perfected by a Parisian optician, Hofman, who, helped by the physicist and astronomer Pierre Janssen, built a new type of 'pocket spectroscope', presented by Janssen to the Paris Academy in 1862. Secchi contacted Janssen and Hofman, Janssen went to Rome with a 'pocket spectroscope', which he adapted to the new equatorial telescope, and the Italian astronomer began systematic observation of spectra.[239] Between 1862 and 1868 he studied the spectra of more of than 4,000 stars; the rewarding outcome was the first spectral classification of the stars of the greatest importance for the subsequent immense developments in astrophysics.[240] Lorenzo Respighi (1824-1889) was engaged in similar advanced research on star spectra; since 1855 he had been Director of Bologna Observatory, and from 1866 Director of the Observatory on the Campidoglio Hill in Rome; he also introduced important optical innovations in the instrument used for spectroscopic observation.[241] In 1871, Secchi, Respighi and Annibale Tacchini (1838-1905), who was later to succeed Secchi as Director of the Observatory of the Collegio Romano, established the Società degli Spetroscopisti italiani. The scientific scope of the Società was extremely advanced: it aimed at co-ordinating the work, and at publishing the results of the Italian observatories concerning the Sun and the new chemical physical knowledge of the stars; the spectroscope was, at the same time, the research instrument and the epistemological banner of this Società, that which gave rise to the Società Astronomica Italiana (1920).[242] In 1872 Tacchini founded one of the first journals of astrophysics in the world, *Giornale degli spettroscopisti italiani*.[243]

Secchi was also a prolific writer. He collected all that was known on the Sun as an astronomical object in a large work, *Le Soleil*, published in 1870 in Paris. The book was a great success and, in 1875, a second edition appeared in three volumes. In *Le Soleil* there was a splendid collection of coloured plates; the same richness of colours could be found in another book, published in Milan in 1877 (*Le stelle. Saggio di astronomia siderale*). In this book, among the many topics addressed, Secchi's star classification had an important place, and there were several pages on the Universe, life and the "Wisdom of the Author of all things", pages that might have been written by Maxwell, in a natural philosopher's style.[244] But Secchi had already written his best-seller in 1864, *L'unità delle forze fisiche*; in this book, he had used the entire knowledge about the mechanical theory of heat, his own researches and those of Melloni on radiant heat, in order to overcome any realist conception of force and of action at a distance. Of course this was the same field already ploughed by Grove, Tyndall, Tait and Maxwell, and the book had four Italian editions, the second in 1874 and the fourth in 1884.[245] At the international level, its success was assured by its inclusion in the prestigious series of the International Scientific Library, a cultural enterprise proposed at the Edinburgh meeting of the British Association for the Advancement of Science, and launched after "the disasters of the war" of 1870, as a joint project with the editors of the French *Revue Scientifique*.[246] Books written by the top specialists in the various topics[247] were published at the same time in French, English, German, Russian and Italian. So, in a period in which positivism had penetrated into Italy principally in the humanities, the image of Italian science was brought to the *"classes éclairées"*[248] of Europe and United States, wearing a cloak of inductive, positivist language, beneath which it is easy to see the black gown of a subtle Jesuit, interested in reducing the philosophical relevance of science, and in subduing inductivism "to a theistic vision and to a Catholic apologetics".[249]

In the section below about the historiography of Italian science - in the wider context of international scientific life - we will see that the quest for the 'precursor' has continued to plague Italian historiography of science, particularly during the Fascist period between the two World Wars. A precursor, in my view, cannot exist; it is a mere artefact of the whiggish approach of history. Nobody can carry on research while (consciously) acting the role of the precursor of someone who, in many cases, is 'waiting' to be born.[250] The figure of the naturalist Agostino Bassi (1773-1856) is dangerously similar to that stereotype. However, while the concept of 'precursor' is better rejected, the status of being a 'pioneer' may be felt subjectively, and - much more important - the knowledge-situation of a 'pioneer' is describable; this evocative metaphor seems legitimate and perhaps useful. Sometimes the two concepts have been freely mixed, as in Bulloch's historical judgement on Bassi: "It is impossible to read the works of Bassi without realising that he was a pioneer and that he was, as *Italians* like Calandruccio, Riquier, Grassi, and Monti have always maintained, the founder of the doctrine of parasitic microbes and a precursor of Schwan, Pasteur, and Koch".[251] The epistemological features of Bassi's enterprise cannot be closely examined here, but some aspects of his work and of the fate of his discoveries display traits of the community of life-scientists that were in striking contrast with those of the astronomers' community, both at the 'national' and international levels. [252]

Like many naturalists, Bassi was not trained in any scientific technique, and never held an academic position; he took a degree in law at Pavia but, meanwhile, like many other students, he was attracted by the presence in that University of the most celebrate *letterati* of the time, Volta and Spallanzani.[253] Spallanzani's lectures must have been really fascinating for a (future) naturalist,[254] all of whose contemporaries positively agree on the professionalism of the great biologist, who was described as a clear and lively teacher.[255] For many years Bassi was a civil servant in Lodi, near Milan, under the French and Austrian régimes; in 1807 he began research on a disease of silk-worms called *muscardine* in France, and *mal del segno, mal del calcino* o *calcinaccio* in many regions of Italy. The disease was very damaging to the economy of whole districts of Lombardy, and many researchers had tried to find some correlation between the breeding conditions and the state of illness of the silk-worms, but without any meaningful results. Bassi, at the beginning of the 1820s, found the right experimental approach: he abandoned he prevalent idea that the disease developed spontaneously as a result of unknown factors, and began to look for a "foreign germ" (*germe estraneo*), believed to be the active, live organism responsible for contagion. After a long and ingenious series of experiments, he found that the *germe* was a cryptogam, later named in his honour *Botrytis bassiana*; finally, in 1832, he was able to isolate and prevent the disease through a particular technique of disinfecting and breeding. In particular, as disinfectants he used simple and easily-available chemicals, such as calcium chloride and potassium nitrate.[256] But, through ill health and bad luck in sheep-breeding, his own circumstances had been severely reduced - indeed at one time he had almost starved;[257] he tried to improve his uncertain economic situation by keeping his findings secret and advertising his disinfectant, but without commercial success.[258] At last he decided to make his results known, and did it in two steps. In 1833 he asked the Imperial Royal University of Pavia for permission to communicate his findings, renewing his request in 1834, when he was able to make his experiments before a mixed commission of professors of the Faculties of Medicine and of Philosophy. In this way he obtained an official certificate, dated 30 August 1834, that confirmed the scientific value of his research. In 1835 he published the first, theoretical, part of the work, and in 1836 the second, practical part.[259] I will trace some of Bassi's ideas through the scientific community.

It is clear that Bassi's conclusions were of interest to a large and varied audience, from agronomists to botanists and physicians. The first reactions, locally and abroad, were almost immediate. A Milanese botanist, G. Balsamo-Crivelli (1800-1874), at once assigned the cryptogam identified to a the genus, and in 1835 published the results in the well-known *Biblioteca italiana*; later, in 1838, he confirmed and extended Bassi's research in a second article, again in the *Biblioteca italiana*. In France, two communications on the silk-worm were presented in 1836 by French authors to the Académie des Sciences, and were quickly published in its *Comptes rendus*, the weekly publication read by the entire European scientific community; both of these French authors confirmed Bassi's findings.[260] Another important fact to be mentioned is that a *résumé* of Bassi's volumes was published in France, as an independent booklet, in 1836.[261] In Germany, Bassi's research became widely known through its key role in Jacob Hanle's *Pathologische Untersuchungen,* 1840. Hanle was an assistant to Johannes Müller in Berlin, and an intimate friend of Theodor Schwann, already famous for his work on fermentation (1837) and cell theory (1839). Thus Hanle disclosed his conjectures on the living origins of contagion from an important centre for the life sciences, and among the few experimental works he was able to quote, Bassi's research on *mal del segno* had a central place.[262] Later, between 1851 and 1854, Carlo Vittadini (1800-1865), a physician and botanist who worked as assistant in the botanical garden of Pavia University, confirmed Bassi's thesis, and made a particularly important advance when he succeeded in growing *Botrytis bassiana* on several substances, such as honey, oil, sugar, and mannitol.[263] These and probably many other printed works form a link between Bassi's *Del mal del segno*, published in Lodi in 1835, and Pasteur's *Études sur la maladie des vers à soie*, published in Paris in 1870, especially considering - at the border with technical literature - the papers and books of agronomists interested in silk-worm breeding.

2.4 Transitions: Journals and Other Problems

The bibliographies of Italian scientists are a good source of evidence about the journals that published their papers. I have chosen three authors whose published papers cover the period 1760-1860; across such a long period, these bibliographies are no more than a thin thread, but they afford a glance through the most important Italian periodicals. The first scientist is Carlo Lodovico Morozzo, who appeared as vice-president of the Accademia Reale delle Scienze in the Royal Patents that established the Academy in 1783. His first paper had been published in volume V (1770-1773) of the *Mélanges de philosophie et de mathématique de la Société Royale de Turin*; Morozzo signed the paper as M^r le comte Mouroux, and all the papers in the *Mélanges* are in French (by Maquer, Monge, La Grange, Condorcet, M^r le comte Saluces) or in Latin (Cigna and Dana). In 1782 he published in the first volume of *Memorie della Società Italiana*, whose birth together with that of Lorgna's Società Italiana was noted above. In the following years Morozzo sent several papers to the Parisian *Journal de Physique*, the last one in the year of his death, 1804. His other papers are divided between *Memorie della Società Italiana delle Scienze* (in Italian) and the *Mémoires de l'Académie Royale des Sciences* (in French), but this latter journal was not published at all between 1793 and 1801, when it reappeared under a new title (*Mémoires de l'Académie des Sciences de Turin*). Morozzo also sent a paper in Latin to the *Acta Bononiensi Instituti* in 1791.[264]

The second scientist I have chosen lived in a cultural milieu completely different from Count Morozzo's austere and bellicose Piedmont, and was entering adulthood as the Restoration was trying to erase all trace of the French convulsions. Bartolomeo Bizio (1791-1862) was by training and profession a pharmacist and lived most of his life in Venice. Bizio's first paper was published in 1827, in the *Giornale di fisica, chimica, storia naturale, medicina ed arti* (1818-1827) then edited by Gaspare Brugnatelli (1795-1852). This journal had its root in the activity of Luigi Valentino Brugnatelli, Gaspare's father, who in 1788 had established *Biblioteca fisica d'Europa* (1788-1791), a journal he continued as *Annali di Chimica* and *Annali di Chimica e Storia Naturale* (1792-1805), and finally as *Giornale di fisica, chimica e Storia naturale* (1808-1818). This last editorial enterprise had an international audience; it was launched when Brugnatelli held the Chair of Chemistry at the University of Pavia (from 1796), and its other editors were Vincenzo Brunacci (1768-1818), holder of the Chair of *matematica sublime* from 1801, and P.Configliachi (1779-1844), Volta's friend and his successor in the Chair of Physics at the same University.[265] Bizio's first paper was also published in the *Opuscoli chimico fisici del farmacista Bartolomeo Bizio* (1827), edited by himself.[266] In that period, Bizio was a well-known chemist and also published in the *Atti dell'Imperial Regio Istituto Veneto*, and in the *Memorie della Società Italiana delle Scienze*.

Of greater significance here are the many papers sent by Bizio to the new *Annali delle Scienze del Regno Lombardo Veneto*, from its first publication onwards. This journal was founded by Ambrogio Fusinieri (1775-1849) in 1830, as a continuation of Brugnatelli's *Giornale*; it lasted until the death of the founder. Fusinieri has an important place in the history of Italian physics, not so much for his scientific activity, as rather for his recurrent criticism of Laplacian physics in favour of an observational and speculative science, more agreeable to the new Romantic mind. Many important scientists working in different fields - mathematics, astronomy, physics and chemistry - contributed to Fusinieri's journal. An analysis of subscribers gives an idea of its distribution: it was read all over the Veneto area, in Milan, Pavia, Modena, and Geneva.[267] In 1843, Bizio published for the first time in the *Annali di fisica, chimica e matematiche* edited by Gianalessandro Majocchi. Majocchi, a physicist and engineer, had graduated from Pavia in 1816, and in the years 1817-1819 had studied at the Vienna Polytechnic.[268] His *Annali* were an outcome of the successful *Riunioni* of Italian scientists, and, following the interests of their editor, they had a strong cultural bias towards topics of applied science and engineering. The sciences were classified in a way that shows an increasing disciplinary specialization: physics, chemistry, mathematics, astronomy, mechanics; alongside original papers, the *Annali di fisica, chimica e matematiche* also published summaries of foreign papers from European technical and scientific periodicals.[269] The differences from Fusinieri's *Annali delle Scienze* were clearcut, reflecting political and other differences between the two editors: in short, Fusinieri was a man of the Restoration,[270] and Majocchi a man of the Risorgimento. This last aspect of Majocchi's engagement is evident in the appeal to Italian scientists published in the fifth volume of his journal (January-March 1842). He urged the collaboration of those authors who - "unfortunately because of vanity" - preferred to publish their results in the French periodical press. They had not understood the role and aim of the yearly *Riunioni* of the scientists, nor did they defend Italian honour. The attack was mainly against the French bias of Italian scientists, and Majocchi's search for a scientific, cultural, and, covertly, political centre is extremely clear. He wrote: those conceited scientists "give arguments to the foreigners by which they may judge Italy unable to support a centre of her knowledge, as regards physical, chemical and mathematical sciences".[271]

Returning for a last time to the quiet, versatile and prolific Bizio,[272] in 1856 he published a paper in *Il Nuovo Cimento, Giornale di fisica, chimica e scienze affini*, a journal which can be discussed in connection with the last of our bibliographic guides, the radical and internationally renowned Calabrian chemist Raffaele Piria. While Bizio only published a few notes abroad, in the French *Comptes rendus* (since 1844) and *Cosmos* (in 1860), Piria published his first important papers when he was in France in the *Annales de Chimie et Physique* and in the *Journal de Chimie Médicale*. After his return to Naples he published in the *Antologia di Scienze naturali*, founded in 1841 by himself and by two other well-known scientists, Arcangelo Scacchi (1810-1863) and Macedonio Melloni. The journal, unfortunately, lived only a year, succumbing to the usual malaise of Italian scientific journals: the dual shortage of subscriptions and of papers.[273] In 1842, Piria moved to Pisa,[274] where he initially published in the local *Miscellanee di chimica, fisica e storia naturale*. In Pisa he found Mossotti and Matteucci, scientists, like himself, of international experience and repute; together they launched a new journal *Il Cimento*, which survived from 1843 until 1847.[275] In the same period, Piria also published in Majocchi's *Annali di fisica, chimica e matematiche* and in a Venetian journal published by the abbé Francesco Zantedeschi, the *Raccolta Fisico-chimica Italiana ossia Raccolta di memorie originali di chimici, fisici e naturalisti italiani*. Zantedeschi edited this collection in the years 1846-1848, and after an understandable pause, published *Annali di Fisica* (1848-1850) in Padua.[276] After a while, Piria and Matteucci, in association with the botanist Giuseppe Meneghini (1811-1889)[277], again returned to editorial labour and in March 1855 published the first issue of the *Nuovo Cimento, giornale di fisica, chimica e scienze naturali*, a journal that - at last! - was able to survive and is still alive today. An analysis of the collaborators shows that the journal was once again local, notwithstanding the ambitions of the promoters and a certain distribution abroad. Piria's last scientific paper was published in the *Nuovo Cimento* in 1856, two extracts from his text-books followed in 1857. After that, and until his death in 1865, Piria was much more interested in politics than in science.

Five years after Piria's death, a group of chemists gathered in Florence, then the capital of the Kingdom of Italy. The meeting was held in the laboratory of Ugo Schiff, where in addition to the host there were only six other chemists (among them Cannizzaro and Selmi). The small group decided to publish the *Gazzetta Chimica Italiana*, and by a lucky coincidence,[278] the meeting in Florence took place just a few days after Rome was takenover by the Piedmontese assault troops, the *bersaglieri*. The chemists' project involved the diffusion throughout the Country, the publication of the best original works of Italian researchers, and the review of several German, French and English journals. The *Gazzetta* was used by Cannizzaro and Schiff as an instrument of the 'chemical colonization' of Italy, and its history shows the up and downs of the many Italian centres of chemical research.[279] The starting of the new chemists' journal had profound consequences for the Italian scientific community, because in a few years the chemists' collaboration with the *Nuovo Cimento* gradually come to a halt. In 1873, the *Nuovo Cimento* began a new series with the sub-title *Giornale di fisica, fisica-matematica, chimica e scienze naturali*.[280] After the death of Piria and Matteucci, the physicist Riccardo Felici (1819-1902) and the mathematician Enrico Betti (1823-1892) became the two editors of the *Nuovo Cimento*; they followed the trend towards specialization, and transformed the journal into the preferred vehicle of scientific communication for Italian physicists.

Thus, after 1870, two of the most representative scientific disciplines had their specialized journals. The natural sciences followed a similar process: the scientific journals progressed from a regional audience to a national one, from an occasional life to a permanent one, from quasi-personal editing to quasi-communal editing, from a 'general-purpose' setting to a disciplinary one. However, one problem remained unsolved: how to participate in the international community using a language of culture such as Italian, which could not become internationally dominant, but was also not so historically marginal (or peripheral) as to be excluded from scientific discourse. The use of languages other than Italian is the clear symptom of this lasting malaise during the period considered. A complete list of important books written by Italian scientists directly in French would be impressive. I will list some of the chief books published in the XIX century by authors quoted in this paper: *Mémoires pour servir à l'histoire naturelle et principalment à l'oryctographie de l'Italie*, by A.Fortis, 1802; the three volumes of *Théorie du mouvement de la lune*, by Plana, 1832; *Sur les forces qui régissent la constitution intérieure des corps*, by Mossotti, 1836; *Traité des phénomènes electro-physiologiques des animaux*, by Matteucci, 1844; *La thermochrôse, ou la coloration calorifique*, by Melloni, 1850; *Cours special sur l'induction, le magnétisme de rotation, le diamagnétisme, et sur les actions moléculaires*, again by Matteucci, 1852; *Principes de Thermodynamique*, by Saint-Robert, 1865; *Le Soleil*, by Secchi, 1870. As a last example of a great, but private, enterprise I quote the work of Stoppani, who, in collaboration with Giuseppe Meneghini and A.Cornalia (1824-1882), published *Paléontologie lombarde ou description des fossiles de Lombardie*, in four volumes between 1859 and 1881; the volumes were printed privately, and Stoppani was helped financially by his father, a small merchant of Lucca. With these volumes, printed after the Unity of Italy, but again in French, I reach the end of this inquiry.

3. PROVISIONAL CONCLUSIONS

The first of the three following sections aims to give a very short review of the past and present approach of Italian historiography to the placing of Italian science in the context of 'science as such', or in relation to the international 'scientific centres'. The second section treats some questions raised by the presence in Italy of a 'natural' powerful centre like the Roman catholic Church. The third and last section of this paper returns to the problem of Dante's bones and gives, as a provisional conclusion, some remarks on the meaning of the inverted commas around the phrases 'science as such' and 'scientific centres'.

3.1 Centre vs. Periphery: Past and Present in Historiography

My intention in this tentative analysis of Italian science historiography is very modest. I wish simply to review the interpretations of some problems of the relationship between Italian science and European science presented in a few past and present works. All the works analysed are collective, with only one exception, that of Icilio Guareschi, who was active at the end of the XIX century and the beginning of the XX century. I have already quoted his essays several times; here I draw attention to his introduction to "Chemistry in Italy from 1750 to 1800".[281] The title of Guareschi's study is a double understatement, because it considers more than 'chemistry' and also covers a much longer time-span. What is fascinating in this essay is the author's intention to study in depth the points of strength and weakness of *Italian culture* of the period, and its relationships with *European culture*. Much emphasis is given to the negative influence of an immense number of clergy, and of Catholicism itself, contrasted with the beneficial influence on scientists of the Reformed Churches. While the attacks on the Roman Catholic Church may - in part - be a heritage of *Risorgimento* traditions, other themes, such as "Italians outside Italy" and "Scientific journeys" are significant and well documented. Guareschi was not haunted by spectral 'precursors'.[282] Guareschi was substantially an *erudito*, and many times he simply collected 'facts'. His approach had a major merit, however: it was (tentatively) holistic, in the sense that it considered many scientists, including the lesser ones; many disciplines, more or less related to chemistry; and many aspects of the concrete activity of the scientists (teaching, institutions, correspondence, political attitudes, journeys, etc.). The extent to which this (tentative) holistic approach has been left out of the working horizon of Italian historiography may be measured by considering five later works, all of them collective.

In 1939, on the occasion of the centenary of the first meeting of Italian scientists, a massive work was published on the history of science in Italy. If one browses through the many volumes of *Un secolo di progresso scientifico italiano (1839-1939)*, edited in the full vigour of the Fascist Regime,[283] the impression is amateurish, inevitable in a collective work in which each speciality is given to a 'contractor' who is always an active scientific researcher, working in the field. The general outcome of the enterprise, as Redondi has written, is the celebration of a fancied 'people of scientists and inventors', [284] and from this point of view the only relation with other 'national' sciences is a competition for priority; some of its contributions have since been used as a statistical source on the scientists active in the different periods.[285] However, in several cases the contributions usefully draw attention to men and events outside the view of historiographically more qualified sources.[286] This aspect of the attention paid to lesser authors is intrinsic to a historiography that celebrates, at one at the same time, a nation and a discipline.[287] It has recently been rediscovered, and interpreted in a more sophisticated way, by Redondi: "the lesser [figures] acquire a remarkable importance for the history of a scientific, localized culture".[288] In the narrative and historiographic structure of *Un secolo di progresso scientifico italiano* science is throughout highly fragmented into specialities, without any attempt at an overall view, not only of science but even of chemistry or physics. A general approach to science was assumed to be impossible, even after the event. In this sense the volumes published in 1939 are 'modern' in conception.[289]

In 1980, the Turin publisher Einaudi published a collection of essays on the history of science and technology in Italy. The editor of the volume, Gianni Micheli, addressed the problem of the 'provincialism' of Italian culture. Micheli discusses different definitions of provincial culture: a fragile and feeble culture, unable to impose its own values on other cultures; a culture that suffers from a sense of inferiority towards other stronger cultures; a culture that is closed within its own boundaries; or one that welcomesindiscriminately whatever arrives from abroad. According to Micheli, since the XVII century, Italian culture has been provincial, but only in the two first senses, because it has always shown the highest degrre of openness to new ideas and themes of research.[290] Micheli's volume, of more than 1,300 pages, was organised according to selected topics, some particular ones, such as the control of waters, and other more general ones, such as the role of science in industrial development. After almost 20 years it remains an indispensable source, with the advantages and the drawbacks of a thematic history, profound on certain points and silent on others.

In the last three initiatives I was myself involved. In 1989, Carlo Maccagni and Paolo Freguglia edited a large volume (1000 pages) aiming to cover the whole history of science in Italy since the XV century. The editors chose - or were compelled - to give very different amounts of space to the individual sciences, so that medicine had 235 pages, life science 120 pages, physics 110 pages and chemistry about 60 pages. On the whole the editing was weak, so that the essays were of unequal historiographic value, even in terms of their critical apparatus; the international context was present in moderate doses. Ferdinando Abbri and I were able to coordinate our respective contributions, but, for example, the physics of the first half of the XIX century was not covered at all. Nevertheless the quantity of information collected in the volume is enormous, and it is particularly rich for the period I am considering in this essay.[291]

In the same year, 1989, another collection of essays was edited, by Vittorio Ancarani, on the history of the scientific disciplines in Italy, with two historiographic aims: to concentrate on institutional aspects and to start the research from Unification.[292] The context of the historical research was explicitly sociological. Many papers - including mine - were discussed in seminars at the Department of Social Sciences of Turin University, then directed by Filippo Barbano. In their introductory papers, Barbano and Ancarani remarked that, in many essays, "the tension between centre and periphery" had been taken as explanatory principle.[293] Here, centre and periphery are understood *inside* Italy, and the prevalence of this interpretive frame confirms that after Unification the Government imposed on Italy a sort of bureaucratic centralism, in pure Piedmontese style. I think it is fair to say that, after a long search for a 'centre,' lasting for two or three generations, Italian scientists found only a central bureaucracy, a meagre result indeed. But Ancarani's volume raises a methodological point of importance in every comparative analysis between the scientific production of two national communities. My colleagues and I, in a large part of our essay in Ancarani's book, developed an analysis of the European context in which the Italian chemical community lived for the three decades after Unification, comparing it quantitatively with the German and English communities.[294] We found to our surprise that the scientific production of the Italian community was quantitatively comparable with that of the English community. This result added interest to our project of a *qualitative comparison* between Italian and English chemistry, and we tried a couple of 'quality controls'.[295] Our results depended critically on the criteria of comparison used; but my point here is that too often a scientific community is said to be more or less advanced than another one, without discussing in depth the criteria used.[296]

In 1992, a workshop on the history of science in Italy was organised by two private Italian groups of historians of mathematics and of chemistry,[297] supported by the prestigious Istituto Italiano di Studi Filosofici in Naples and by the "Renato Caccioppoli" Department of Mathematics of the University of Naples. In preliminary discussion, the organizers decided to concentrate the analysis on the period around the unification of Italy, and to call for papers by researchers open to methodological debate. Moreover, the organizers decided on a provoking (and hopeful) title: "a history yet to be made".[298] In order to give a scenario within which many contributions could be placed, it was also decided to propose a synthesis of the interactions between physics and chemistry in the period; unexpectedly the difficulty of connecting 'local' events with what was happening at the 'global' level became evident.[299] Later, Simonetta Di Sieno edited the papers in two instalments;[300] but, participants had the immediate impression at the end of the workshop that the evident cleavage in the political history of the country (Unification) did not correspond to so clear a fracture in the history of science. Albeit provisional, it was a major result, which opened many questions, of course still open.

All these attempts, and several others that I have not mentioned, have suffered from the isolation of the history of science in the context of Italian culture. Apparently it is possible to write a very long essay on "culture" from Unification until 1970 without any, even tentative, analysis of the 'function' of scientists and engineers in the context of the so-called learned society.[301] If this can happen in a 'progressive' historiographic context, it is easy to imagine the fate of the history of science in a context more directly connected with the tradition of Benedetto Croce (1866-1952). Though Croce has had a negligible impact on the philosophy of the XX century[302] his presence in Italian culture is still cumbersome. Speaking on the shortcomings of the Italian history of science, Paolo Casini quoted Croce's epistemological condemnation of science: "the pretence of a mathematical science of nature is inadmissible"; Casini also recalled that Croce in his *Logica come scienza del concetto puro* (1909), laid the debate on the fundaments of mathematics to rest with a few witticisms.[303] Casini's view of the negative effect of Croce's literary gospel is surely right, but it has always been met by a symmetrical philosophical weakness on the part of scientists. Roberto Maiocchi has recently questioned the attitude of Italian scientists towards the philosophical problems of science, just at the time of Croce's *Logica*, and has argued their almost total indifference.[304]

The indifference of Italian contemporary culture towards the cultural meaning of science and technology has very deep roots. Since the trial of Galileo, science and technology have unceasingly and progressively been shut away in a sort of epistemological ghetto. The influence of the Catholic Church on an understanding of science as *public knowledge* has been heavy and negative. Pietro Redondi has underlined that "in order to establish an agreement between science and faith, the

Catholic Church has always insisted that it accepts science insofar as it is a method".[305] The best thing that scientists could do was a good, possibly a sophisticated experiment, without any theoretical dream (or philosophical nightmare). In other words, science, as result and cause of a complex cultural, historical process, could not be an important component of a *Weltanschauung*.

3.2 An Inner Cultural and Political Centre: the Roman Catholic Church

> *Tractabitur Logica, Physica, Metaphysica, Moralis scientia et etiam Mathematicae, quatenus tamen ad finem nobis propositum conveniunt.*
> Jesuits' *Ratio studiorum*, 1599.[306]

Since Galileo, in the Italian cultural context, scientists have limited the objective knowledge of Nature to the 'direct', experimental manifestation of facts, distrusting the theoretical, conceptual and systematic sides of science. This attitude was common to many of the scientists whom we have met in this enquiry. Lazzaro Spallanzani was probably the most innovative among the XVIII century life scientists; he performed an enormous number of experiments, however he refuted to give a general account of his method and - owing to this anti-system bias - this task was left to his Genevan friend Jean Senebier (1784).[307] Two generations later the Italian best theoretical physicist, Mossotti, stated that "phenomena themselves are not explained" (*i fenomeni propriamente non si spiegano*); there are no theories, but only more general facts, or common principles; what is obtained by physics is "to see how other individual facts derive from the existence of that general fact". Mossotti started from the perspective of French positivism, but was unable to give science the wide social and cultural meaning given by the French positivists.[308] Whole scientific community, like the chemists' one, have discarded any serious theoretical interest for a century, from Cannizzaro to Giulio Natta, so that theoretical contributions appear as exceptions to a constrained rule.[309] In his long essay Luigi Besana supports by many documents the permanance in Italian culture of a purely utilitarian attitude towards science, and shows how that attitude has always led to teach science by precepts.[310]

The generality and permanence of this unconcern on the cultural value of science has deep roots in Italian history; here we may focus our attention on the educational system, and on the lasting predominance of the Roman Catholic Church in the education of the Italian ruling class, all over the period here considered. In this way, and at the same time, we will obtain a *partial* answer to several other questions raised in my narrative. The first question regards the high number of researchers active in Italy, under the (mostly) unstable and restrictive conditions here described. With few exceptions all the quoted scientists were connected with teaching in Universities or military Academies; from this point of view the division of the country in several states has supported a number of scientific institutions which had a counterpart only in the German states. At the moment of the Unification the new-born state had to harmonise laws and by-laws of sixteen state Universities and four free (*libere*) Universities.[311] Almost all these Universities had a long tradition, and the successful example of the reform of the Universities of Turin and Pavia had been followed in many other cases; in particular the salary of the professors had been improved, as well as the endowments of the Chairs of Natural and Physical Sciences and of Medicine.[312] Sometimes, as we have seen above, the competition between the princes favoured the social status of scientists.

In presenting the many characters acting on the Italian stage I have always tried to point out their social origin, and in particular if they were aristocrats or members of the Catholic clergy. On the important role of aristocracy in Italian culture I commented in section 2.2, the role of clergymen in the scientific development deserves a few general remarks, and a more articulated analysis in the case of the Society of Jesus. The Italian educational system was dominated by the Catholic Church, so many University Professors were priests or members of the learned orders (*Barnabiti*, *Scolopi*, Jesuits, etc.); in almost all the cases these Professors were perfectly integrated in the life and culture of their University, without any duty about the cure of souls, so, from a sociological point of view, their over-all scientific relevance is completely similar to that of their lay colleagues.

The discourse about the Jesuit Order has to be deepened, because of its internationalism, its commitment to education, and - finally - to its structural role in the situation mapped in this enquiry. The internationalism of the Society is strictly pertinent to our story because the role of Jesuits in the life of the different Italian states followed the ups and downs of the Society in the European context - including its disbanding in 1773 and reconstitution in 1814. About the Jesuits' particular commitment to education, and its results in the Italian situation, I prefer to call upon members of the Catholic clergy or of the Catholic laity, who lived in the period considered here, and had, on their own account, an important role in Italian culture.

Antonio Gentili has considered the life of Paolo Frisi from the point of view of his being a *Barnabita*; in this religious context Gentili points out the contrast between Frisi ("a man of enlightment") and the Society of Jesus, and in particular Frisi's attacks to the Jesuits' schools. Frisi cosidered the schools of the Society "dark and foggy" (*oscure e caliginose*); in that schools "the young's' subordination was asked for, more than their sound education", and the Jesuits' aim was "a kind of command (*impero*) on opinions and business of men".[313] These judgements of Frisi's were collected and published in 1787. After sixty years the wish of another important priest was tremendous: "to uproot the rotten and incorrigible sect, so that no relict abides on the earth". This course may be read in Vincenzo Gioberti's *Il gesuita moderno*, a ponderous work in five volumes published in Lausanne in 1846-47.[314] In another book, the *Prolegomeni del Primato* (1845), Gioberti remarked an aspect of Jesuits' teaching and policy which hints to the core of the relationships between the Order and modern, experimental science: to alienate "the public progress from the knowledge (*cognizioni*) which brings about it, it is a reviling attack to God, repugnant to the order and aim of world, baleful for men and against the essence, concepts and purpose of Christianity".[315] As we have seen above (section 1.3), Gioberti's thought about the cultural and religious roots of the Italian primacy (*Primato*) had been published in 1843, in a book that deeply influenced the whole political thought of the *Risorgimento*, particularly through the adhesion to many Gioberti's ideas by many Piedmontese moderate Catholics. The success of Gioberti's thought in the moderate part of the Piedmontese ruling class was "great and quick", because the *Prolegomeni* and other books of Gioberti's gave "a new cultural basis to the hope of renewing the alliance between throne and altar as an alliance between civilisation and religion".[316] Among the diffusers of Gioberti's books in Turin there was a young student of hydraulic engineering, Quintino Sella (1827-1884), a Catholic with a Jansenistic shade,[317] who will became a good mineralogist and an important statesman after the Unification. In a letter to his brother (September 1847) he comments the nomination of two new Ministers by the King in these terms: "S.Marzano is of feeble character, has two sons in the Jesuit Order [...] Broglia is of staunch character, has a son in the Jesuit Order, and dislikes all the ideas of moderns".[318] But the direct influence of the Jesuits on the Piedmontese policy was coming to the end: the moderate Catholics formed the ruling class of Piedmont from the 1848 revolution until the Unification (after the Unification were an essential part of the Italian ruling class), so it is not surprising that between the end of 1848 and the beginning of 1849 Gioberti himself was Prime Minister of the King Carlo Alberto.

Members of the Catholic clergy like Frisi and Gioberti attacked mainly the Jesuits' educational method, but the concern of the Catholic laity was often wider, and involved the entire influence of the Church in the Italian schools. Niccolò Tommaseo (1802-1874) was one of the most important Italian writers of the XIX century; fervent Catholic believer, he was Minister of Education in the revolutionary Government of Venice in 1848-49. In 1831 he published a book on the political and religious situation in Italy, and on the innovations which he believed necessary for the country (*Delle innovazioni religiose e politiche buone per l'Italia*). In the book he condemned "the violent and exclusive influence of clergy on the education of laity [...], the cure of ministers of cult in inculcating a blind indefinite submission to sovereigns [...], the intolerance towards diverse or adverse religions".[319] While Tommaseo was Minister in Venice, Giuseppe Montanelli (see above, section 1.3) was one of the three leaders of the Provisional Government in Tuscan. He had passed from Gioberti's side to the democratic[320] side, and after the defeat of the revolution he openly asked the end of the Pope's temporal power. In a book published in Turin in 1851, Montanelli connected the Pope-King, in power in Rome, to "the Pope-King of Russia and Rothschild"; in other terms the problem of the temporal power was an international problem, Italians were compelled to an "*inner struggle* of freedom against the Pope, and because of the cosmopolitan character of the adverse force, the *inner conflict* involve[d] a European conflict".[321] I have emphasised the terms 'inner struggle' and 'inner conflict' because that was the situation not only in Montanelli's political and geographical context, but also in the conscience of many Italian Catholics.

Sella, by then an eminent statesman of the Right, was Finance Minister in 1862, 1865 and from 1869 until 1873, thus he was in the Italian Government at the moment of the annexation of Rome. For him, in 1848, the "reactionary party" was "represented by Austria and Jesuits"; in 1876, after the death of the Bishop of Biella, a friend of his, he depicts the new Bishop like "a blind instrument of Jesuits".[322] Sella was really uneasy in his threefold role of Catholic, scientist and politician, and for him the relationship between Science and Faith was "a white-hot topic" for decades.[323] Guido Quazza has published parts of a moving manuscript by Sella - simple notes without syntactic connection - about this relationship; the notes were probably written in 1876. He was "concerned increasing estrangement Science and Religion"; it was necessary that the Pope wins his fear: "Does not fear science progress - that fear I believe has commanded provisions made to prevail by Jesuit faction against convictions prelates more enlightened".[324]

From Frisi until Sella, many Italian Catholic scientists saw the Society of Jesus completely submitted to its own *ratio studiorum*; they openly admitted: the study of any science is permitted to us *ad finem nobis propositum,* to the aim planed by us. The aim could be noble, but the instrumental character of the Jesuits' interest in science produced a specific kind of experimental research and of science teaching, both - research and teaching - abundant but fragmented, in order to attenuate, enfeeble, manipulate the cultural impact of science.[325]

3.3 Centre vs. Periphery, or: neither Centre, nor Periphery

> Ahi Pisa, vituperio delle genti
> del bel paese là dove 'l sì sona
> Dante Alighieri, La Divina Commedia
> Inf. XXXIII, vv. 79-80

At the end of a terrible description of the death by starvation of Count Ugolino and his four sons, Dante curses Pisa, hoping that all its inhabitants might die by drowning in the river Arno, the same river that flows through Florence; the poet *must* imagine an apocalyptic disaster, because the peoples living near Pisa had not yet punished the city. The two lines of the epigraph are simply the beginning of the curse, where Dante says that Pisa is the shame of the people living in the Country where *sì* resounds. We are interested in two different aspects of this *incipit*. The first one is that this malediction strikes a town less than a hundred kilometres from Florence, Dante's birthplace; in effect, Dante's curses against Italian towns are numerous, and reflect a Country divided into a myriad city states, endlessly fighting for local or very local supremacy. Dante, the 'national' Italian poet, had no notion of Italy as a potentially single State. The second interesting aspect is that Dante gave a *linguistic definition* of Italy, but the poet considered Italian, French and Provençal as three different ways of pronunciation of the same tongue. From Dante's times until now the problem of multi-linguism has been always relevant, sometime fundamental, in Italian culture.[326] In this sense the difficulties of Italian scientists after the giving up of Latin as international language were by no means new.

The introductory volume of an imposing history of Italy has a very simple title: *L'Italia come problema storiografico* [Italy as historiographic problem]. One of the questions addressed by its editor, Giuseppe Galasso, is the fundamental "distinction between political consciousness and linguistic-literary awareness",[327] a distinction that has permitted the lasting existence of Italians without any existing Italy. If Galasso sees Italy in Dante's time developing into a "modern multi-ethnic and centrifugal nation",[328] in the period considered here the centrifugal forces were overcome, Italy became a national State, but the presence in the Country of many striving cultural and political centres was plainly confirmed. The polycentric nature of Italian culture brings with it richness and extravagance. More specifically, in the case of scientific culture the many stories we have followed permit us a few considerations, but before that I want to look at the whole historical process that brought to the Unity of Italy.

The Unification of Italy may seem not much less than a historical miracle, if the continuous, pervasive, heavy influence that the Holy See had in the political and religious life of the Peninsula is considered. In this kind of 'miracle', the weight of events and processes of European relevance was overwhelming: on one hand the impact of the French Revolution and of the Napoleonic invasion, on the other hand the dramatic loss of political and cultural power of the Roman Catholic Church. Of course that power loss had become evident under the pressure of 'enlightened rulers' such as Maria Theresa and Joseph II. However, the Revolution and the Napoleonic wars had a profound, if not a univocal, effect on the 'national' conscience of the Italian aristocracy and bourgeoisie, at least by encouraging the perception that the European political setting was mobile, and changeable in favour of Italian Unity. Against such a background, the loss of prestige and of political power of the

Church is understandable. It became ruinous after 1848-49, when it was clear that the Holy See was *against* the Unity of Italy. Even so, at the end of the XVIII century the Catholic clergy was still able to mobilise the populace, as the tragic death of hundreds of Neapolitan Jacobins showed the world. But this type of violent defence of the established 'order' was no longer possible in the second part of the XIX century. Many years ago, Antonio Gramsci (1891-1937) affirmed that the ability of the liberal movement to stir up the "liberal-Catholic strength" as its ally was "the political masterpiece of the *Risorgimento*".[329] In the particular story told here, we have seen the professor Domenico Cirillo, the physician and botanist, hunged in 1799 to the cry of *Viva Maria*, and in 1870 professor Angelo Secchi, Jesuit and astronomer, collaborating with the lay Government of unified Italy. Within one life span, the relationship between the Holy See and the Italian bourgeoisie changed in an eloquent way.

The fading of the Roman Catholic centre was a major factor permitting Italian scientists to look for new, and possibly their own, centres of reference for political support, professional legitimisation, and social visibility. However, the several actual centres, dispersed through the many Italian States, were obliged to follow the will of their rulers. Pavia depended on Vienna, and therefore on the general policy of the Imperial Chancellery towards the periphery of the Austrian Empire (a quite specific centre-periphery dialectics). On one hand Turin followed the fate of the Savoy dynasty, and, on the other, it was at all times under strong pressure from the pervasive French culture. In a similar way, University of Pisa and the Florentine institutions were supported by the dynastic role and ambitions of the Lorrainese Grand-Dukes. Unfortunately for Italian science, when at last geography declared that there was only one Italian State, the centre in Rome was fundamentally a bureaucratic centre, in the pure old Piedmontese style.

One may wonder whether other institutions, in a less physical space, could have been better centres for Italian science. I refer to the attempts embodied in Lorgna's Società Italiana delle Scienze and in the *Riunioni* of the scientists. The marginal role of the Società was confirmed by the *Riunioni* themselves, which were planned and held outside it. The function of the meetings was more complex: they were used instrumentally by the political power of the Italian States in evident reciprocal competition, but they were also perfectly integrated into a European scenario. For the scientists, the *Riunioni* also transmitted consciousness and knowledge. The scientists acquired a greater awareness of the realities and potential of their profession, and during long stays in different parts of Italy they improved their knowledge of the social, economic and cultural conditions of the Country.[330]

Until now I have considered the scientists as members of a *scientific community*, structured by common political, professional and institutional interests, and I have shown that the structure was loose and had many centres. But when we consider other analytic categories, such as the *disciplines*, we are in danger of observing the past through the very selective filters of the disciplinary histories, and of being deceived by the legitimating function of these histories. In this research I have made much use of this kind of history; certainly the final result suffers from the same defects as do the sources. However, for the general aim of inquiring into the centre/periphery dialectics, we may study the actual procedures that, in any discipline, are considered to lead to scientific knowledge, or, in brief, the *knowledge procedures* actually applied by scientists. These procedures have many components (which I have discussed elsewhere); [331] a short list would include: (1) the criteria of demarcation, e.g. between science and non-science, backward and advanced science, pure and applied science, research inside and outside the discipline; (2) theories, their 'spare parts', and the criteria for their use; (3) observational and experimental practices; (4) instruments, chemicals, machines and plants, including the norms for their sound use; (5) views about the stability of disciplinary knowledge; (6) closure criteria (when is research or a research field 'closed'?); (7) disciplinary languages and speciality dialects. Depending on its importance in any particular context, each component may induce the scientist to look to a particular 'centre'. I cannot discuss every constituent of the knowledge procedure in detail, but I will try to give some examples of the actual working of the first four elements.

The criteria of *demarcation* depend to a great extent on assumed levels of prestige, utility, and so on. Piedmontese scientists had a French style in distinguishing between pure and applied science; the chemists participating in the *Riunioni* had very local criteria for the choice of topics of advanced research. Probably *theories* and their use are the component of the knowledge procedure that historians have paid most attention to, even to absurdity, often extracting only an abstract formalised frame from the actual epistemic structure in use. When Plana and Carlini were working on the theory of the movements of the Moon, they constructed particular 'spare parts' to replace the semi-empirical methods then applied in the theory of the Moon. For that work, the 'centre' was somewhere in a 'discussion space' between Turin and Milan. When Fortis was classifying plants and shells, Piazzi was observing Cerere, Spallanzani was blinding bats, and Volta was constructing electrical devices, all four *letterati* were making observations and doing experiments in the style of XVIII century science, but each was using a particular, personal and disciplinary set of rules, actions and material devices. *Observations* and *experiments* are so different - even synchronically - that to speak of a common (methodological) centre is plainly impossible.

A different discourse concerns *instruments, chemicals,* and *machines,* because it is possible to examine the arrival of orders from the peripheries, and the shipment of goods from the centres of production. It is also possible to follow the scientists' journeys to Vienna, Paris and London, in search of balances, optical instruments and chemicals. In this sense, Paris and London were real centres for the international community, as *markets of scientific goods*. It is to be remembered that the practical use of even apparently simple instruments often requires the transmission of tacit knowledge, that has to be learned on the spot, working together with experts.[332] But it is also to be noted that *laboratories* were not (and still are not) necessary for all scientific disciplines, nor were they open to students for a large part of the period considered here. After the 1820s, the necessity and accessibility of laboratories were accepted by the disciplinary communities, and by then *international schools* had been born in chemistry, physics, physiology, and bacteriology - probably in that order. A school may be a *real* centre, if 'real' means 'occurring in fact'. For a historian, this means the presence of documents or artefacts by means of which he or she may infer that something was 'real'. In this vein I have studied a *Tischgesellschaft* of Heidelberg, of which, in 1856, Agostino Frapolli and Angelo Pavesi (1830-1896) were members, together with L.Meyer, Kekulé, Landolt, Roscoe, Beilstein, and many others.[333] It can be shown that in the period considered here, the concentration of researchers, means and ideas in Italy never reached the level of the French and German schools, even in the case of Pavia at the end of the XVIII century or of Pisa in the 1840s. Just as these Universities were approaching that goal, political events disrupted the dawning schools. It was really a pity, because styles of studying, teaching and living were spread throughout the learned world by such centres; it is perhaps for this reason that schools are so often the model used in sociology or in history to identify relationships between centre and periphery. From this point of view, the Italian chemical community depended on Paris in the 1840s and 1850s, but after that a transition of interest towards a constellation of Universities in the German *koiné* began.

The lasting question of language deserves a final comment. Among the components of knowledge procedures, I listed disciplinary languages and speciality dialects, but I do not intend here to investigate more deeply the cognitive value of linguistic devices such as the 'dense' Latin of botanists or the structural formulae of organic chemists. In the context of the present research, I am referring only to natural languages as means of communication between scientists. Here the *kind* of international relevance of a natural language is of paramount importance. In an epoch of diffuse Anglomania in Italy, Matteucci married an English girl; a man of religious attitude, he was also greatly influenced by Newton's version of the relationship between science and religion,[334] but nevertheless he found it better to publish his books in French. From the end of the XVIII century until the second half of the XIX century, the French scientific journals were the preferred channel of communication for Italian scientists; after 1850, in increasing numbers they sent their papers to the German *Zeitschriften*. It is fair to say that in the XIX century scientific Italian was a provincial language.

In conclusion, if science is interpreted as a purely formal, logical and hierarchical activity, then it may be simplified, with one or more geographical centres that supply peripheries with rules for correct research and for up-to-date teaching. But a different perspective is opened by the shift from an abstract disembodied science to a science understood as the outcome of real scientists' activities. In a simple metaphor, the change is from a geometry that considers distances from a geographical or epistemological centre, to a topology that looks for relations, only slowly affected by the continuous change of shape of the historical phenomena. From this last viewpoint, we can say that the Italian scientific community was and remained polycentric in many fundamental aspects of scientific activity, and thus it was neither centre nor periphery. A more peripheral role was played in crucial sectors, however, depending on the production of certain instruments and chemicals, on the experimental training of the practitioners of some specialities, and on quick and efficacious communication to the international community. If one looks only to the markets for goods, training and information that characterized the activities of some specialities, then Italy was peripheral to these markets. If adequate attention is directed to the whole assortment of disciplines and specialities that we call 'science', the centre/periphery opposition cannot be applied at all.

Dipartimento di Chimica Generale ed Organica Applicata, Università di Torino, Italy

NOTES

[1] Quoted from: C.Dionisotti, *Geografia e storia della letteratura italiana*, Torino: Einaudi, 1967, p. 28. The revision of his great history occupied Tiraboschi from 1787 until his death.

[2] D. Della Terza, "Le Storie della letteratura italiana: premesse erudite e verifiche ideologiche," in: A.Asor Rosa (ed.), *Letteratura italiana*, vol. IV, *L'interpretazione*, Torino: Einaudi, 1985, p. 311-349, on p. 311. The edition of Tiraboschi's work studied by Dante Della Terza was: G.Tiraboschi, *Storia della letteratura italiana*, 8 tomes in 16 volumes, published in Milan between 1822 and 1826.

[3] C.Dionisotti, *Geografia e storia della letteratura italiana*, cit., on p. 29.

[4] *Dissertazione preliminare sull'origine del decadimento delle scienze*; on the relationship between Tiraboschi's *Dissertazione* and the *Reflections critique* of the abbé Du Bos (1719), see: D. Della Terza, "Le Storie della letteratura italiana," cit., p. 312.

[5] E.Raimondi, "Letteratura e scienza nella «Storia» del Tiraboschi," in: R.Cremante, W.Tega, *Scienza e letteratura nella cultura italiana del settecento*, Bologna: Mulino, 1984, p. 295-309, on p. 302-305. Raimondi underlines that Tiraboschi's discourse begins with a reference to a contribution of R.Boscovich, another famous Jesuit to whom many references will be made in this research.

[6] A modern critic has spoken of a "dreary landscape" in many literary genres; see: D. Della Terza, "Le Storie della letteratura italiana," cit., p. 318.

[7] The words *toscana favella* were printed in italics in the original text. "Rinunzia avanti Nodaro degli Autori del presente Foglio periodico al vocabolario della Crusca" was the ironic title. The Accademia della Crusca was the major institution dedicated to the defence of XIV century Italian, as used by Boccaccio; see: B.Migliorini, *Storia della lingua italiana*, Firenze: Sansoni, 1978, p. 509-512.

[8] Quoted from: B.Migliorini, *Storia della lingua italiana*, cit., p. 515.

[9] G.Galeani Napione, *Trattato dell'uso e dei pregi della lingua italiana*, Torino, 1791; see: G.Devoto, *Profilo di storia linguistica italiana*, Firenze: Nuova Italia, 1976, p. 112-113.

[10] B.Migliorini, *Storia della lingua italiana*, cit., p. 499.

[11] D.Mattalia (ed.), *La divina commedia. Inferno*, Milano: Rizzoli, 1993, p. XXXI.

[12] In a great part the history of Dante's tomb is obscure, but the name of this prudent friar is known because he wrote on the urn: *DANTIS OSSA a me Frate Antonio Santi hic posita.*

[13] G.A.Scartazzini, N.Scarano, *Dantologia. Vita e opere di Dante Alighieri*, Milano Hoepli, 1894, p. 203; D.Mattalia (ed.), *La divina commedia. Inferno*, cit., p. LXVIII.

[14] G.A.Scartazzini, N.Scarano, *Dantologia*, op. cit., p. 200.

[15] F.Chabod, *L'idea di nazione*, Bari: Laterza, 1974, p. 52.

[16] Ib., p. 65.

[17] W.Binni, "Vittorio Alfieri," in: E.Cecchi, N.Sapegno (eds.), *Storia della Letteratura Italiana*, vol. VI, *Il Settecento*, Milano: Garzanti, 1968, p. 907-1024, on p. 1014.

[18] B.Migliorini, *Storia della lingua italiana*, cit., p. 609. In the rich landscape of the Italian Romanticism Manzoni may be singled out because he joined an impassioned adhesion to the Unity ideal to a deep commitment to a personal religious search within the Roman Catholic Church; see: G.Alberti, "Alessandro Manzoni," in: E.Cecchi, N.Sapegno (eds.), *Storia della Letteratura Italiana*, vol. VII, *L'Ottocento*, Milano: Garzanti, 1969, pp. 619-745, on pp. 650 and 658-661.

[19] Ib., p. 593; italics in Foscolo's text.

[20] G.Devoto, *Profilo di storia linguistica italiana*, Firenze: Nuova Italia, 1976, p. 101-114. French could be also the *private* language, as in the case of the mineralogist Giorgio Santi that published books in Italian, but wrote his diary in French; see: B.Migliorini, *Storia della lingua italiana*, cit., p. 507 and 526.

[21] G.Devoto, *Profilo di storia linguistica italiana*, cit., p. 105.

[22] B.Migliorini, *Storia della lingua italiana*, cit., p. 525.

[23] M.Borsa, *Del gusto presente in letteratura italiana*, Venezia, 1784; quoted from B.Migliorini, *Storia della lingua italiana*, loc. cit.

[24] G.Devoto, *Profilo di storia linguistica italiana*, cit., p. 110-111.

[25] B.Migliorini, *Storia della lingua italiana*, cit., p. 502; I stress the *Verdinglichung* of the servant people.

[26] Ib., p. 619.

[27] Ib., p. 525.

[28] Ib., p. 619.

[29] Ib., p. 594.

[30] It is very important to note that a part of the *Italian literature* was (and is) written in the different regional dialects, and that the dialects need to be *translated* in Italian; see: P.V.Mengaldo, *Poeti italiani del Novecento*, Milano: Mondadori, 1981. In this anthology the translations often are by the same 'dialectal' poets. At any time the 'literary tongue' has been a sort of middling Italian, and many writers used in the cultured prose words taken from the *local dialects*. In the period here considered the same Tuscan writers used words

taken not only from the spoken Florentine, but also from the tongues spoken in Arezzo, or in Leighorn, or in small area as the Mugello Valley; see: B.Migliorini, *Storia della lingua italiana*, cit., pp. 648-652.

[31] T.De Mauro, *Storia linguistica dell'Italia unita*, Bari: Laterza, 1963; this book has been and still is a fundamental and seminal research on the real use of dialects in Italy, and Devoto, while quoting the book, remarks that it confirms "the unpopularity and the oligarchic nature of the Italian literary tongue"; see: G.Devoto, *Profilo di storia linguistica italiana*, cit., p. 179.

[32] F.Chabod, *L'idea di nazione*, cit., p. 67; italics in the text.

[33] B.Migliorini, *Storia della lingua italiana*, cit., p. 591.

[34] M.Di Bono, "L'astronomia in Italia dal Quattrocento alla prima metà del Novecento," in: C.Maccagni, P.Freguglia (eds.), *Storia sociale e culturale dell'Italia*, vol. V, *La cultura filosofica e scientifica*, t. II, *La storia delle scienze*, Busto Arsizio: Bramante, p. 3-57, on p. 43. The condemnation of Copernicanism lasted from 5 March 1616 until 25 September 1822; see: *Scienziati e tecnologi* (later ST), Milano: Modadori, 1975, vol. III, p. 626. The question of the *Index* will be touched on later, now it is enough to recall that it was an instrument of religious integralism and political control; as such it was not a particularly medieval institution, and this is confirmed by the date of its establishment, in 1564, by Pope Paul IV.

[35] *The Oxford Encyclopaedic English Dictionary*, Oxford: Oxford UP, 1991, s.v.; my italics.

[36] G.Bollati, "L'italiano," in: R.Romano, C.Vivanti (ed.), *Storia d'Italia*, vol. 1, *I caratteri originari*, Torino: Einaudi, 1972, p. 951-1022; quoted on p. 967.

[37] Ib. p. 958.

[38] A.Fortis, *Saggio d'osservazione sopra l'isole di Cherso e Osero*, Venezia, 1771, quoted from: F.Venturi, *Settecento riformatore*, vol. V, *L'Italia dei Lumi*, t.II, *La repubblica di Venezia (1761-1797)*, Torino: Einaudi, 1990, p. 415.

[39] Ib., p. 415.

[40] G.Paoli, *Ruggiero Giuseppe Boscovich nella scienza e nella storia del '700*, Roma: Accademia dei XL, 1988, p. 21.

[41] Ib. p. 302.

[42] F.Venturi, *La repubblica di Venezia (1761-1797)*, cit., p. 417.

[43] Ib. p. 418.

[44] Quoted from: G.Paoli, *Ruggiero Giuseppe Boscovich*, cit., p. 510.

[45] G.Tagliaferri, "Boscovich and Milan," in: M.Bossi, P.Tucci (eds.), *Bicentennial commemoration of R.G.Boscovich*, Milan: Unicopli, 1988, p. 9-20, on p. 14.

[46] G.Paoli, *Ruggiero Giuseppe Boscovich*, cit., p. 288-291, 511.

[47] Paoli discusses the question at many points of his book, in particular on p. 499-513. For an extensive bibliography see p. 561-565.

[48] A.Genocchi, "Luigi Lagrange," in: *Il primo secolo della R.Accademia delle Scienze di Torino, 1783-1883*, Torino: Paravia, 1883, p. 86-95, on p. 93. Angelo Genocchi (1817-1899) was a Turinese mathematician.

[49] Ib., p. 92.

[50] Ib., p. 94.

[51] L.Pepe, "Lagrange e i suoi biografi," in: F.Burzio, *Lagrange*, Torino: UTET, 1993(1942), p. XI-XLIII, on p. XIV.

[52] J.L.Lagrange, *Mècanique analytique*, Paris: Courcier, 1811.

[53] L.Pepe, "Lagrange e i suoi biografi," cit., p. XVI.

[54] He died near Chambery, in a territory that then belonged to France; see: L.Bulferetti, "Menabrea, Luigi Federico," ST, s.v.

[55] I willreturn to these meetings in section 2.2.

[56] D.Palladino, "La matematica italiana dall'inizio dell'Ottocento alla seconda guerra mondiale," in: C.Maccagni, P.Freguglia (eds.), La storia delle scienze, cit., p. 216-261, on p. 224-225.

[57] M.Galluzzi, "Geometria algebrica e logica tra Otto e Novecento," in: G.Micheli (ed.), Scienza e tecnica nella cultura e nella società dal Rinascimento a oggi, Torino: Einaudi, 1980, p. 1003-1105, on p. 1028-1029.

[58] Piedmont was plainly a province of the French Empire; see: L.Cerruti, "Amedeo Avogadro," in: Tra Società e Scienza. 200 anni di storia dell'Accademia delle Scienze di Torino, Torino: Allemandi, 1988, p. 132-137, on p. 133.

[59] A.Avogadro, "Essai d'une manière de déterminer les masses relatives des molécules élémentaire des corps et les proportions selon lesquelles elles entrent dans ces combinaisons ," Journal de Physique, d'Histoire naturelle et des Arts, 73, p. 58-76 (1811).

[60] L.Cerruti, Uomini e idee della chimica classica, Milano: Eurobase, 1985, p. 28-30. Here I refer to Dumas, Gaudin and Regnault as members of Ampère's invisible school.

[61] L.Cerruti, "Il luogo del Sunto," in: S.Cannizzaro, Sunto di un corso di filosofia chimica, Palermo: Sellerio, 1991, p. 77-282, on p. 94-95.

[62] A.Gaudiano, "Malaguti, Faustino Giovita Mariano," ST, s.v.

[63] G.Provenzal, Profili bio-bibliografici di chimici italiani, Roma, s.d. (but 1938), p. 156.

[64] M.U.Dianzani, "Le scuole mediche chirurgiche," in: F.Traniello (ed.), L'Università di Torino. Profilo storico e istituzionale, Torino: Pluriverso, 1993, p. 93-110, on p.100.

[65] R.Foraggiana, A.Dohrn, "Dohrn, Anton," ST, s.v.; M. Alippi Cappelletti, "La biologia italiana dell'Ottocento," in: C.Maccagni, P.Freguglia (eds.), La storia delle scienze, cit., p. 492-533, on p. 522-523.

[66] G.Groeben, "Dohrn, Anton F.," Dizionario Biografico degli Italiani (later DBI), s.v.; on the role of Naples' zoological Station in the development of biology see: A.Monroy, C.Groeben, "La stazione zoologica di Napoli ed il suo ruolo nello sviluppo della biologia," in: P.Nastasi (ed.), Il Meridione e le Scienze, Palermo: Istituto Gramsci Siciliano, 1988, p. 29-38. I am sure that the DBI will publish (in many years) also Moleschott's biography.

[67] Quoted from: F.Diaz, "Politici e ideologi," in: E.Cecchi, N.Sapegno (eds.), Storia della Letteratura Italiana. Il Settecento, cit., pp. 57-306, on p. 143.

[68] M.De Zan, "La messa all'Indice del «Newtonianismo per le dame» di Francesco Algarotti," in: R.Cremante, W.Tega (eds.), Scienza e letteratura nella cultura italiana del settecento, cit., pp. 133-147, pp. 134-138.

[69] A.Rupert Hall, "La matematica, Newton e la letteratura," in: R.Cremante, W.Tega (eds.), Scienza e letteratura nella cultura italiana del settecento, cit., pp. 29-46, on p. 37.

[70] M.De Zan, "La messa all'Indice del «Newtonianismo per le dame» di Francesco Algarotti," cit., on p. 134.

[71] The title page of the 1758 edition was illustrated by a scene of book-burning; the epigraph was a quotation from the Gospel: Multi eorum qui fuerant curiosa sectati, contulerunt Libros, et combusserunt coram omnibus, Act., XIX, 19; see: A.Rotondò, "La censura ecclessiastica e la cultura," in: R.Romano, C.Vivanti (ed.), Storia d'Italia, vol. 5, I documenti, Torino: Einaudi, 1973, pp. 1397-1492, on p.1398.

[72] Op. cit., p. 1487.

[73] The French translation was published in Paris in 1766, with a false edition place (Philadelphia); the Greek translation was also published in Paris (1802). Meanwhile *Dei delitti e delle pene* had been traslated in English (London, 1767), Spanish (Madrid, 1774), and it was finally to be translated into Russian and printed in Petersburg in 1803; see: F.Diaz, "Politici e ideologi," cit. , pp. 179-183.

[74] A.Rotondò, "La censura ecclesiastica e la cultura," cit., p. 1491.

[75] This is how Voltaire referred to Needham in private corrispondence and in his publications; see: J.Rostand, *Lazzaro Spallanzani e le origini della biologia sperimentale*, Torino: Einaudi, 1963, p. 42.

[76] The book was printed in Modena, and the essay against spontaneus generation was published with another in Latin (*De lapidus ab acqua resilientibus*). Bibliography in: M.L.Altieri Biagi, B.Basile (eds.), *Scienziati del Settecento*, Milano-Napoli: Ricciardi, 1983, pp. 171-176.

[77] W.Bernardi, "Scienze della vita e materialismo nel Settecento," in: P.Rossi (ed.), *Storia della scienza moderna e contemporanea*, vol. I, *Dalla Rivoluzione Scientifica alla Età dei Lumi*, Torino: UTET, 1988, pp. 567- 590, on p. 573.

[78] Voltaire was happy to accuse Needham of potential atheism; see: J.Rostand, *Lazzaro Spallanzani e le origini della biologia sperimentale*, cit., p. 42-43.

[79] C.Castellani, "Spallanzani, Lazzaro," ST, s.v.

[80] Similar statements may be found in: P.Redondi, "Cultura e scienza dall'illuminismo al positivismo," in: G.Micheli, *Scienza e tecnica nella cultura e nella società dal Rinascimento a oggi*, cit., pp. 679-811, p. 683. This point will be discussed in relationship to the Italian historiografy of science in section 3.1.

[81] P.Casini, *Newton e la coscienza europea*, Bologna: Mulino, 1983, p. 160. In the second part of the *Dissertazionis de Lumine* the chronological time of the lecture is fully recorded: *die 5 septembris hora 2*; evidently day and hour felt important by the author; ib., p. 169.

[82] Ib., p. 170.

[83] A.Wandruszka, "Il mondo politico europeo nel XVIII secolo," in: G.Mann, A.Nitschke (eds.), *I propilei*. vol. VII, *Dalla riforma all'illuminismo*, Milano: Mondadori, 1973, pp. 445-532, on p. 518.

[84] Firmian was born in Trento, and his father was an adviser to Maria Theresa. His political career was wholly spent in the Austrian diplomatic service, and he represented the Vienna Government in Milan for 23 years.

[85] For a large analysis of Frisi's life and work see: G.Barbarisi (ed.), *Ideologia e scienza nell'opera di Paolo Frisi (1728-1784)*, Milano: Angeli, 1987, two volumes.

[86] A.Gigli Berzolari, *Alessandro Volta e la cultura scientifica e tecnologica tra '700 e '800*, cit., on pp. 145 and 354. Volta was a friend of Georg Christof Lichtenberg, professor of physics in Göttingen, and protagonist of German enlightment.

[87] E.Brambilla, "Le professioni scientifico-tecniche a Milano e la riforma dei collegi privilegiati (sec XVII - 1770)," in: G.Barbarisi (ed.), *Ideologia e scienza nell'opera di Paolo Frisi (1728-1784)*, cit., pp. 345- 446, on pp. 389-390 and 438. The two clergy's attitude towards the lumi was deeply different, as well as their appreciation of the mètaphysical 'content' of mechanics and mathematics. Frisi supported a mechanics oriented towards a rigorous application to practical problems, Boscovich - as it is well known - was the principal proponent in the XVIII century of a dynamic theory with a metaphysical background; on this divergence, which led to explicit polemic between Frisi and Boscovich see:- P.Redondi, "Cultura e scienza dall'illuminismo al positivismo," cit., pp. 685-698.

[88] Ib., pp. 149-150.

[89] P.Vaccari, Storia dell'Università di Pavia, Pavia: Università di Pavia, 1957, pp. 164-211. In this part of Vaccari's book there is a rich documentation about the buildings and the other structures that enriched the University of Pavia in the period of the reforms. Particulary suggestive the plate from Scopoli's Deliciae florae et faunae insubricae (1786), where it is depicted the botanical garden; on p. 167.

[90] Ib. pp. 190-191. Later Pollack designed also a splendid 'villa' for Bonaparte in Milan.

[91] M.Gliozzi, "Beccaria, Giovanni Battista," ST, s.v.

[92] P.Nastasi, "Paolo Frisi e il problema della forma della terra," in: G.Barbarisi (ed.), Ideologia e scienza nell'opera di Paolo Frisi (1728-1784), cit., vol. I, pp. 99-144, on pp. 107-108, 128.

[93] V. de Alfaro, "Gli studi di Fisica," in: F.Traniello (ed.), L'Università di Torino. Profilo storico e istituzionale, cit., pp. 227-235, on pp. 227-228.

[94] A.Gigli Berzolari, Alessandro Volta e la cultura scientifica e tecnologica tra '700 e '800, cit., p. 213.

[95] A fine plate from Michelotti's Sperimenti idraulici (1767) is in: V.Marchis, G.Jarre, "Accademici o Tecnologi?," in: Tra Società e Scienza. 200 anni di storia dell'Accademia delle Scienze di Torino, cit., pp. 92-107, on pp. 98-99. In this essay the two authors demonstate the Turin academicians' constant interest in the application of scientific knowledge to economic development.

[96] M.Gliozzi, "Bidone, Giorgio," ST, s.v.; see also: P.Redondi, "Cultura e scienza dall'illuminismo al positivismo," cit., on pp. 770-773.

[97] G.Paoloni, "Scienza, università e accademie dagli Stati preunitari allo Stato unitario," in: S.Di Sieno (ed.), Scienze in Italia, 1840-1880. Una storia da fare. Parte I, Quaderni PRISTEM, n 4, 1993, pp. 1-32, on p. 13.

[98] F.Venturi, "L'Italia fuori d'Italia," in: R.Romano, C.Vivanti (eds.), Storia d'Italia. Dal primo settecento all'Unità, cit., pp. 987-1481; on p. 1042.

[99] On this institution, and on the rich - in every sense - culture that fed it, numerous contributions may be found in: R.Cremante, W.Tega (eds.), Scienza e letteratura nella cultura italiana del settecento, cit.; see also: M.Cavazza, Settecento inquieto. Alle origini dell'Istituto delle Scienze di Bologna, Bologna: Mulino, 1990.

[100] M.Di Bono, "L'astronomia in Italia dal Quattrocento alla prima metà del Novecento," cit., pp. 31-32.

[101] Ib., p. 33.

[102] This exemplary Theoretical Physics professor was the President of the first meeting of the Italian scientists, as Decane of the University of Pisa, in 1839; see: I.Cantù, L'Italia scientifica contemporanea, notizie sugli italiani ascritti ai cinque primi congressi, attinte alle fonti più autentiche, Milano: Stella, 1844, s.v.; on this book vide infra, and later in section 2.2.

[103] E.Miotto, G.Tagliaferri, P.Tucci, La strumentazione nella storia dell'Osservatorio astronomico di Brera, Milano: Unicopli, 1989, pp. 12-15. Boscovich assigned to these funds his salary as Professor of the University of Pavia; see: E.Brambilla, "Le professioni scientifico-tecniche a Milano e la riforma dei collegi privilegiati (sec XVII - 1770)," cit., p. 438.

[104] M.Di Bono, "L'astronomia in Italia dal Quattrocento alla prima metà del Novecento," cit., p. 35.

[105] It has been anlysed in: M.Pera, La rana ambigua. La controversia sull'elettricità animale tra Galvani e Volta, Torino: Einaudi, 1986.

[106] J.L.Heilbron, "The contribution of Bologna to Galvanism," Hist.Stud.Phys.Scie., 22, pp.57-85 (1991).

[107] F.Blezza, Galvani e Volta: la polemica sull'elettricità, Brescia: La Scuola, 1983, pp. 54-58; G.Pancaldi, "Electricity and life. Volta's path to the battery," Hist.Stud.Phys.Scie., 21, pp. 123-160 (1990).

[108] C.Castellani, L.Usuelli, "Galvani, Luigi," ST, s.v.

[109] The judgement of Italian historians on this period is deeply divided, because it depends on other very controversial assessments concerning the French Revolution and the process that led to the Unification of Italy, the *Risorgimento*. I personally feel ethical and religious revulsion towards men like Napoleon, whom I consider plain slaughterers.

[110] S.J.Woolf, "La storia politica e sociale," in: : R.Romano, C.Vivanti (ed.), Storia d'Italia, vol. 3, Dal primo settecento all'Unità, Torino: Einaudi, 1973, pp. 5-508, on p. 192.

[111] A.Gigli Berzolari, Alessandro Volta e la cultura scientifica e tecnologica tra '700 e '800, cit., p. 194.

[112] Also Moscati and Scarpa were followers of Brown; see: P.Redondi, "Cultura e scienza dall'illuminismo al positivismo," p. 703.

[113] I.Guareschi, "La chimica in Italia dal 1750 al 1800," Suppl. Ann. Enc. Chim., 25, pp. 327-378 (1908-09), on p. 353.

[114] I.Guareschi, "Giovanni Fabbroni," Suppl. Ann. Enc. Chim., 25, pp. 449-464 (1908-09), on pp. 451-453.

[115] I.Guareschi, "Gaspare Ferdinando Felice Fontana," Suppl. Ann. Enc. Chim., 25, pp. 411-448 (1908-09), on p. 416.

[116] G.Panseri, "La nascita della polizia medica," in: G.Micheli (ed.), Scienza e tecnica nella cultura e nella società dal Rinascimento a oggi, cit., pp. 157-196, on pp. 193-194.

[117] G.Bilancioni: "Domenico Cotugno," in: A.Mieli (ed.), Gli scienziati italiani dall'inizio del medio evo ai giorni nostri, Roma: Nardecchia, vol. I, 1921, pp. 164-183, on p. 169.

[118] C'est dans son arrière-boutique que se tenaient, à l'abri des sbires, les mysterieuses assemblées des parisans de la liberté (P.Pisani, 1893), quoted from: V.Giormani, "Vincenzo Dandolo, uno speziale illuminato nella Venezia dell'ultimo '700," Ateneo Veneto, 175, pp. 59-130 (1988), on p. 98.

[119] Ib., pp. 99-104. Dandolo and other Jacobeans found a provisional refuge in the Cisalpine Republic.

[120] F.Chabod, L'idea di nazione, cit., p. 59; the phrase originates with P.R.Rohden.

[121] In an anonymous contemporary engraving we see the wagons slowly rolling along in a romantic landscape; each wagon is hauled by five pairs of oxen; see: E.Cecchi, N.Sapegno (eds.), Storia della Letteratura Italiana, vol. VII, L' Ottocento, Milano: Garzanti, 1969, p. 257; among the many treasures on the wagons were paintings by Raphael and Titian.

[122] From Pavia sixty volumes of Haller's herbarium were removed; see: P.Vaccari, Storia dell'Università di Pavia, cit., pp. 219-220.

[123] The italics are Guareschi's; see: I.Guareschi, "La chimica in Italia dal 1750 al 1800," cit., pp. 348-351.

[124] S.J.Woolf, "La storia politica e sociale," cit., p. 205.

[125] A.Gigli Berzolari, Alessandro Volta e la cultura scientifica e tecnologica tra '700 e '800, cit., p. 200.

[126] G.Penso, Scienziati italiani e Unità d'Italia. Storia dell'Accademia Nazionale dei XL, Roma: Accademia dei XL, 1978, pp. 151-154.

[127] A.Bassani, "I chimici dell'Istituto Veneto di Scienze Lettere ed Arti in epoca austriaca. - 1840-1866," in: S.Di Sieno (ed.), Scienze in Italia, 1840-1880. Una storia da fare. Parte II,

Quaderni PRISTEM, n 5, 1994, pp. 97-121, on p. 99-100; the quotation is from the article 121 of the Constitution.

[128] The contemporary military historians estimated 450.000 Italians enrolled in the Army and in other armed services, in 15 years of French occupation. Since 1799 until 1814 Piedmont gave 72.000 soldiers; see: F.A.Pinelli, Storia militare del Piemonte in continuazione di quella del Saluzzo, cioè dalla pace di Aquisgrana sino ai dì nostri, vol. II, Dal 1796 al 1831, Torino: Degiorgis, 1854, pp. 379-380. Only in January 1813 a supplementary conscription enrolled 7,720 Piedmonteses; see: F.A.Pinelli, Storia militare del Piemonte. Supplemento ai volumi I e II, Torino: Degiorgis, 1855, p. 14.

[129] It was not simple rhetoric, because the 'French' Army lost the 63% of the solders; D.G.Chandler, Le campagne di Napoleone, Milano: Rizzoli, 1968, p. 1315

[130] I.Cantù, L'Italia scientifica contemporanea, cit.; on Bartolomeo Panizza (1785-1867), the physician, and on Camillo Vacani, the major-general, see sub voce. Vacani's Storia delle campagne was published in Milan (1825, 3 volumes in 4°) and in Florence (1827, 6 volumes in 8°).

[131] Chabod quotes Giuseppe Mazzini (1805-1872) and Vincenzo Gioberti (1801-1852); see: F.Chabod, L'idea di nazione, cit., p. 26.

[132] Associazione Italiana di Chimica Generale e Applicata, Stanislao Cannizzaro. Scritti vari e lettere inedite nel centenario della nascita, Roma: Leonardo da Vinci, 1926, p.338.

[133] L.Salvatorelli, Sommario della storia d'Italia, Einaudi: 1974 (1955), p. 467. Salvatorelli's volume is a classic of Italian historiography.

[134] N.Nada, "la Restaurazione," in: F.Traniello (ed.), L'Università di Torino. Profilo storico e istituzionale, cit., pp. 35-39.

[135] L.Cerruti, "Amedeo Avogadro," in: F.Traniello (ed.), L'Università di Torino. Profilo storico e istituzionale, cit., pp. 335-37.

[136] On the scientific work of Nobili see: A.Gigli Berzolari, Alessandro Volta e la cultura scientifica e tecnologica tra '700 e '800, cit., p. 332 with bibliography and passim. Nobili is strictly connected with the Italian school of electro-physiology, which among other included C.Matteucci (see later). On this school see: R.Taton, Histoire générale des sciences, t. III, vol. I, Le XIXe siécle, Paris: Presse universitaire de France, 1981, pp. 478-479.

[137] The three celebrated scientists examined Melloni's earlier work, required the experiments to be repeated before them and, after calculations and discussions, published a very long, favourable report. For unknown reasons Melloni's mémoir was never published; see: M.Gliozzi, "Melloni, Macedonio," ST, s.v. For the general context of research of Melloni see: S.Nunziante Cesaro, E.Torracca, "Correlazioni tra proprietà fisiche e struttura delle molecole: le origini della spettroscopia nell'infrarosso," in: P.Antoniotti, L.Cerruti, Atti del I° Convegno di Storia della Chimica, Torino: Univercittà, 1986, pp. 154-163; see also the present writer's editorial remarks, in which Melloni's 'central' experiment is described.

[138] Ciro Menotti, leader of the Modena revolt, was hanged. He had been the shipping-agent of the Società Italiana delle Scienze, which had been established in 1782 with patriotic intentions, but by then was quietly residente in Modena; see: G.Penso, Scienziati italiani e Unità d'Italia. Storia dell'Accademia Nazionale dei XL, Roma: Accademia dei XL, 1978, pp. 263-265. On this Society vide infra in this section, and specially in section 2.1.

[139] On Cattaneo, the Politecnico and Colombani see: P.Redondi, "Cultura e scienza dall'illuminismo al positivismo," in: G.Micheli, Scienza e tecnica nella cultura e nella società dal Rinascimento a oggi, cit., pp. 679-811, on pp. 733-763.

[140] He had already been Secretary of State in the years 1800-1806, when he had stipulated the concordat with Napoleon.

[141] Quoted from: P.Redondi, "Cultura e scienza dall'illuminismo al positivismo," cit., p. 791.

[142] Ib., p. 796. It is well known that, in military manoeuvres, the rear-guard may abruptly become the avant-guard. After fifty years, at the end of the XIX century, this change in position of Italian culture was favoured by the defeat of positivism at the hands of Croce's idealism, so that, after the concordat with Mussolini, this very Roman Catholic interpretation of science was able to triumph

[143] ST, vol. III, p. 639; E.Bellone, I modelli e la concezione del mondo nella fisica moderna da Laplace a Bohr, Milano: Feltrinelli, 1973, pp.112-113, 261-263.

[144] A.Mottana, "Lo sviluppo delle scienze della terra in Italia nel periodo 1840-1880," in: S.Di Sieno (ed.), Scienze in Italia, 1840-1880. Una storia da fare. Parte I, cit., pp. 33-62, on p. 35, 54.

[145] Two points must be remarked: the unusual professional situation of the physicist, who was the Director of the pharmacy of the Hospital of S.Maria della Croce in Ravenna, and the fact that Ravenna was one of the hundreds of small towns that spotted the map of the Peninsula.

[146] G.Moruzzi, "Matteucci, Carlo," ST, s.v; C.Pighetti, Carlo Matteucci e il Risorgimento scientifico, cit., p. 8.

[147] L.Cerruti, "Chimica e chimici in Italia. 1820-1970," in: C.Maccagni, P.Freguglia (eds), La storia delle scienze, cit., pp. 411-440, on pp. 416-417.

[148] C.Matteucci, Discorso sul metodo razionale scientifico, Forlì: Casali, 1835.

[149] C.Pighetti, Carlo Matteucci e il Risorgimento scientifico, cit., p. 13.

[150] Loc.cit.

[151] Of 27 death sentences 12 were executed. The general that led the repression was rewarded by Carlo Alberto with the highest decoration of the Kingdom; see: L.Salvatorelli, Sommario della storia d'Italia, cit., pp. 424-425.

[152] Ib., p.434.

[153] E.Passerin d'Entreves, "Ideologie del Risorgimento," in: E.Cecchi, N.Sapegno (eds.), Storia della Letteratura Italiana. L' Ottocento, cit., pp. 201-413, on pp. 346-349.

[154] The Piedmontese politician discussed the riots in Romagna, a district of the State of the Church whose capital was Ravenna. In 1843 an insurrection was attempted in Bologna, and in the following repression, seven rebels were put to death; see: L.Salvatorelli, Sommario della storia d'Italia, cit., pp. 425-426.

[155] In my family, it is proudly remembered that three Cerruti brothers fought against the Austrians on the Milanese barricades. Later, all three Cerrutis joined the diplomatic service of the Kingdom of Sardinia; see: Ministero per gli affari esteri, Annuario diplomatico del Regno d'Italia, Torino: Paravia, 1865, pp. 110-111.

[156] Patriotic declarations and dinastic interests were mixed in the flag of the Piedmontese troops, which at the center of the tricolour of old Napoleonic memory bore an enormous emblem of the Savoy dynasty.

[157] See at the end of this paper a comment about Bunsen's school in Heidelberg.

[158] M.B.D'Ambrosio, "Oronzo Gabriele Costa," DBI, s.v.

[159] Many of the information about pharmacists is in: G.Maggioni, C.Masino, A.Russo, Dizionario Storico Biografico dei Farmacisti Italiani, Torino: Academia Italiana di Storia della Farmacia, 1984.

[160] De Vico was a member of the Società Italiana delle Scienze, and an astronomer of the Observatory in the Collegio Romano; see: G.Penso, Scienziati italiani e Unità d'Italia, cit.,

p. 265. Secchi was then a teacher of physics and mathematics; on his important scientific contributions vide infra, section 2.3.

[161] F.Venturi, "L'Italia fuori d'Italia," cit., on pp. 1335-1337. Matteucci's appeal was immediately published in French in Frankfurt.

[162] F.A.Pinelli, Storia militare del Piemonte in continuazione di quella del Saluzzo, cioè dalla pace di Aquisgrana sino ai dì nostri, vol. III, Dal 1831 al 1850, Torino: Degiorgis, 1855, pp. 423-425.

[163] G.Penso, Scienziati italiani e Unità d'Italia, cit., pp. 264-265.

[164] L.Cerruti, "Il luogo del Sunto," in: S.Cannizzaro, Sunto di un corso di filosofia chimica, Palermo: Sellerio, 1991, pp. 77-282, on pp. 203-207.

[165] G.Amoretti, "Cavalli, Giovanni," ST, s.v.; the first barrels were made by Cavalli in Sweden.

[166] P. de Saint-Robert, Principes de Thermodynamique, Turin: Cassone, 1865; chapter VII has the title Mouvement des designiles dans les armes à feu.

[167] A.Drago, "Forza ed energia: l'analisi critica di Saint-Robert," in: S.Di Sieno (ed.), Scienze in Italia, 1840-1880. Una storia da fare, cit., pp. 91-110.

[168] On the scientific and technological innovations triggered by this design see: P.Redondi, "Cultura e scienza dall'illuminismo al positivismo," cit., pp.777-782.

[169] G.Penso, *Scienziati italiani e Unità d'Italia* cit., pp. 9-28.

[170] Ib., p. 58.

[171] Ib., p. 52; capitals in the text.

[172] Ib., p. 44; capitals in the text.

[173] L.Grugnetti, "Aspetti della matematica italiana tra i secoli XVII e XIX," in: C.Maccagni, P.Freguglia (eds.), La storia delle scienze, cit., pp. 193-215.

[174] G.Penso, *Scienziati italiani e Unità d'Italia*, cit., p. 29.

[175] Ib., p. 32.

[176] Ib., p. 44.

[177] He collaborated on the project of an *Enciclopedia Italiana*, of which a *Prodromo* [Prospectus] was published in Siena in 1779. This *Enciclopedia* responded to the wish to produce an Italian text that could tackle the problems of the Italian culture, and that might present the new culture and the new science being produced in Italy; but unfortunately the project aborted due to the untimely death of its principal author, the (now) almost unknown Abbot Alessandro Zorzi; see: E.Raimondi, "Letteratura e scienza nella «Storia» del Tiraboschi," cit., p. 298.

[178] Ib., pp. 34-35.

[179] On p. 38 Penso refers the answers of Volta, Spallanzani, Lagrange, Landriani, Boscovich, Toaldo, Cagnoli, Ximenez, Riccati, Paoli, Saluzzo; however Paoli, Toaldo and Cagnoli were associated later, in 1782, 1791 and 1783 respectively.

[180] Ib., p. 54. Article IX of the Society Statutes (1786) stated that "The Papers of the Members must be written in the Italian tongue (*lingua Italiana*)," capitals in the text, ib., p. 451.

[181] Ib., p. 54.

[182] C.L.Morozzo, "Sperienze sopra il precipitato porpora ottenuto dal gas ricavato dallo stagno e dalla sua calce," Mem.Soc.Ital., **1**, pp. 431-443 (1782). Morozzo held a conservative opinion about flogiston and this paper, curious because of its linguistic fate, was traslated into French by a M. Best of Dijon, and published in Paris: "Expériences sur la pourpre mineral obtenu par le moyen du gaz tiré de l'étain et de sa chause," Journ. Phys., **27**, pp. 241-269 (1785).

[183] Ib., pp. 59-60.

[184] M.Gliozzi, "Venturi, Giovanni Battista," ST, s.v.

[185] Canterzani had worked with Zanotti, but after the death of Zanotti, the direction of Bologna Observatory was given to the old P.Matteucci (1708-1800); Canterzani succeeded Zanotti to the Chair of astronomy; see: M.Di Bono, "L'astronomia in Italia dal Quattrocento alla prima metà del Novecento," cit., p. 32.

[186] G.Armocida, "La medicina e i medici in Italia dal XVI secolo ai giorni nostri," in: C.Maccagni, P.Freguglia (eds.), *La storia delle scienze*, cit., pp. 711-825, on p. 750.

[187] We will see that this general pattern needs an important addition, in regard to Sicily; on the general problem of the sciences in southern Italy in the period here considered see: P.Nastasi (ed.), *Atti del Convegno Il Meridione e le Scienze*, Palermo: Istituto Gramsci Siciliano, 1988.

[188] G.Penso, *Scienziati italiani e Unità d'Italia*, cit., p. 56.

[189] M.Boas Hall, "La scienza italiana vista dalla Royal Society," in: R.Cremante, W.Tega (eds.), *Scienza e letteratura nella cultura italiana del settecento*, cit., pp. 47-64.

[190] G.Penso, *Scienziati italiani e Unità d'Italia*, cit., p. 65.

[191] "The Congresses of Italian scientists [...] had a dual effect: 1) to gather together intellectuals of the highest level, concentrating them and multiplying their influence; 2) to bring about a more rapid concentration and a more decided stance in the intellectuals of the lower levels". From the nineteenth *Quaderno dal carcere*; see A.Gramsci, *Il Risorgimento*, Roma: Editori Riuniti, 1975, p. 130-131.

[192] For a correct critique of this biased attitude see M.Galluzzi, "Geometria algebrica e logica tra Otto e Novecento," cit., fn. 57 above, on p. 1020.

[193] F.Chabod, *L'idea di nazione*, cit., p. 35.

[194] ST, vol. III, p. 624. In 1819 Oken had lost for political reasons his Chair of Medicine at the Jena University; see F.Mondella, "Oken, Lorenz," ST, s.v.

[195] J.Morrell, A.Thackray, *Gentlemen of Science. Early years of the British Association for the Advancement of Science*, Oxford: Clarendon Press, 1981.

[196] I.Cantù, *L'Italia scientifica contemporanea, notizie sugli italiani ascritti ai cinque primi congressi, attinte alle fonti più autentiche*, Milano: Stella, 1844; this book was published in instalments with separate page numbering, on occasion of the VI meeting in Milan; the names of the participants in the first five congesses are given in alphabetical order; see: "A.De Caumont," s.v.

[197] B.Bertini, *Relazione del XIV Congresso scientifico francese tenutosi in Marsiglia nel settembre 1846*, Torino: Mussano, 1847, p. 3.

[198] On this aspect see: G.Pancaldi, "Scientific Internationalism and the British Association," in: R.MacLeod, P.Collins (eds.), *The Parliament of Science. The British Association for the Advancement of Science, 1831-1881*, Northwood: Science Reviews, 1981, pp. 145-169.

[199] W.H.Brock, "Science Education," in: R.C.Olby, G.N.Cantor, M.J.S.Hodge (eds.), *Companion to the History of Modern Science*, pp. 946-959; on pp. 947-948.

[200] The circular is reproduced in: G.Penso, *Scienziati italiani e Unità d'Italia*, cit., p. 251. The addressees were the University professors of scientific disciplines - including medicine, as we have seen - the Heads of the Army Corps of Engineering of the various Italian States, of the Botanical Gardens anf of the Museums of Natural History. In addition, it was sent to eight Italian Academies and "beyond the Alps" (*oltremonti*) to the most famous foreign Academies.

[201] M.Galluzzi, "Geometria algebrica e logica tra Otto e Novecento," cit., p. 1021.

[202] B.Bertini, *Relazione del XIV Congresso scientifico francese tenutosi in Marsiglia nel settembre 1846*, op. cit.

[203] Alippi Cappelletti is somewhat critical of the scientific value of the meetings in regards to the life sciences and reports some contemporary criticisms; see: M. Alippi Cappelletti, "La biologia italiana dell' Ottocento," cit., p. 494.

[204] Ib., pp. 1-3.

[205] J.E.Pétrequin, *Revue médico-chirurgicale des travaux du dixième congrès scientifique de France sept-oct 1842*, Lyon: Imprimerie de Marle, 1842; pp. 3, 32.

[206] L.Giacardi, A.Conte, "Gli studi matematici," in: F.Traniello (ed.), *L'Università di Torino. Profilo storico e istituzionale*, Torino: Pluriverso, 1993, pp. 208-224, on p. 214.

[207] F.Muniz, A.M.Tagliavini, "Identikit degli scienziati a congresso," in: G.Pancaldi (ed.), *I congressi degli scienziati italiani nell'età del positivismo*, Bologna: CLUEB, 1983, pp. 153-170.

[208] When I was young, I was fascinated by the cool poetry of the *Canti*, and by the even cooler prose of the *Operette*. Only after the political backlash of the 1980s was I able to appreciate *I promessi sposi* as a realistic portrait of the national character of the Italians. This autobiographical note has the historiographical intention of bearing witness to the tenacity of certain 'local' values over the centuries.

[209] N.Sapegno, "Giacomo Leopardi," in: E.Cecchi, N.Sapegno, *Storia della Letteratura Italiana. L'Ottocento*, cit., pp. 817-958, on p. 817.

[210] I will return to this important point in the conclusions.

[211] L.Cerruti, E.Torracca, "Developments of Chemistry in Italy, 1840-1910," in press.

[212] P. Passerin d'Entreves, "L'Accademia delle Scienze di Torino e l'Evoluzionismo," in: *Tra Società e Scienza. 200 anni di storia dell'Accademia delle scienze di Torino*, Torino: Allemandi, 1988, pp. 148-157, on pp. 149-150; M. Alippi Cappelletti, "La biologia italiana dell'Ottocento," cit., on pp. 515-516.

[213] M.Galluzzi, "Geometria algebrica e logica tra Otto e Novecento," cit., p. 1024.

[214] Ib., pp. 1023-1033.

[215] E.Torracca, L.Cerruti, "Developments of Chemistry in Italy, 1840-1910," op. cit.

[216] L.Giacardi, A.Conte, "Gli studi matematici," cit., p. 214.

[217] It may be added that the competition among sovereigns helped to attract many famous foreign names; cfr.: G.Pancaldi, "Scientific Internationalism and the British Association," cit., on p. 158.

[218] After writing this part of my article I found - to my great satisfaction, and as confirmation that there is a sort of objectivity in historical research - a fine paper by Edoardo Proverbio, who, in a less impressionistic way (i.e. statistically) arrives at the same structure of Italian astronomy in terms of four generations. (E.Proverbio, "Sui carteggi degli astronomi italiani: per un catalogo della corrispondenza di Angelo Secchi," in: A.Rossi (ed.), *Atti del XIII Congresso Nazionale di Storia della Fisica*, Lecce: Conte, 1995, pp. 309-319.)

[219] P.Maffei, "Piazzi, Giuseppe," ST, s.v.

[220] K.Hudson, "Ramsden, Jesse," ST, s.v.

[221] A picture of the instrument was published by Piazzi in his description of the new observatory (*Della Specola Astronomica*, 1792); see: M.Joffe, *La conquista delle stelle*, Milano: Mondadori, 1958, p. 502.

[222] Ib., pp. 484-485, and p. 503.

[223] J.E.Bode, *Verzeichnis der geraden Aufsteigung und der Abweichung von 5505 Sternen nach den Beobachtungen des Herrn Dr. Piazzi in Palermo*, Berlin: Lange, 1805.

[224] P.Maffei, "Plana, Giovanni Antonio, Amedeo," ST, s.v.; in 1817 he married Alessandra Lagrange, a niece of the great mathematician.

[225] A.Mandrino, G.Tagliaferri, P.Tucci, *Catalogo della corrispondenza degli Astronomi di Brera, 1729-1799*, Milano: Università di Milano, 1986, p. 403.

[226] E.Miotto, G.Tagliaferri, P.Tucci, *La strumentazione nella storia dell'Osservatorio astronomico di Brera*, cit., p. 21.

[227] G.Plana, *Théorie du mouvement de la lune*, Turin: Imprimerie Royale, 1832; the three volumes are in quarto. With very few exceptions for occasional papers, Plana used French in all the numerous papers published on the *Memorie* of the Turin Academy in the years 1811-1868; see: *Il primo secolo della R.Accademia delle Scienze di Torino, 1783-1883*, Torino: Paravia, 1883, pp. 540-543.

[228] G.Plana, "Mémoire sur différents points relatifs à la Théorie des perturbations des planètes exposé dans la *Mécanique céleste*," *Astron. Soc. London*, 2, pp. 325-412 (1826).

[229] D.Palladino, "La matematica italiana dall'inizio dell'Ottocento alla seconda guerra mondiale," in: C.Maccagni, P.Freguglia (eds.), *La storia delle scienze*, cit., pp. 216-261, on p.221; U.Bottazzini, "La «moderna analisi»," in: P.Rossi (ed.), *Storia della scienza moderna e contemporanea*, vol. II, *Dall'età romantica alla società industriale*, t. I, pp. 117-147, p. 135.

[230] The parallel passes through the Alps near the Moncenisio pass.

[231] P.Maffei, "Plana, Giovanni Antonio Amedeo," cit., p. 538.

[232] M.Di Bono, "L'astronomia in Italia dal Quattrocento alla prima metà del Novecento," cit., p. 41.

[233] D.Galletto, "Il contributo dell'Accademia allo sviluppo della matematica e della fisica matematica," in: *Tra Società e Scienza. 200 anni di storia dell'Accademia delle Scienze di Torino*, Torino: Allemandi, 1988, pp. 172-179, on p. 174. Galletto remarks that Plana's scientific work had not been studied from the historiographic point of view.

[234] U.Bottazzini, "La «moderna analisi»," cit., pp. 136-137.

[235] P.Maffei, "Secchi, Angelo," ST, s.v.

[236] P.Redondi, "Cultura e scienza dall'illuminismo al positivismo," cit., on p. 799; the group of scientists was led by mathematician Francesco Brioschi and mineralogist and politician Quintino Sella (on Sella see later, section 3.2).

[237] Loc. cit.; I will return in the conclusions to the complex meaning of this statement, with which I wholly agree.

[238] P.Maffei, "Secchi, Angelo," cit.; Maffei has published a picture of the apparatus: ST, vol. III, p. 98.

[239] S.Mancuso, "P. Angelo Secchi e la spettroscopia stellare," in: F:Bevilacqua (ed.), *Atti del XII Congresso Nazionale di Storia della Fisica*, Milano: CNR, 1994, pp. 219-223.

[240] P.Maffei, "Secchi, Angelo," cit.

[241] M.Di Bono, "L'astronomia in Italia dal Quattrocento alla prima metà del Novecento," cit., p. 44.

[242] ST, vol. III, p. 695.

[243] M.Joffe, *La conquista delle stelle*, cit., p. 568.

[244] Ib., pp. 625-628.

[245] P.Redondi, "Cultura e scienza dall'illuminismo al positivismo," cit., pp. 800-803.

[246] I am reading from a leaflet of the publisher bound with the text of A.Wurtz, *La théorie atomique*, Paris: Baillière, 1879.

[247] For example by Norman Lockyer on spectral analysis, or by Herbert Spencer on social science. Before the publication of Secchi's book, Saint-Robert's *La nature de la force* had already seen its second edition.

[248] In the leaflet already quoted.

[249] P.Redondi, "Cultura e scienza dall'illuminismo al positivismo," cit., p. 809.

[250] The use of expressions like "she was ahead of her times" carries a similar paradoxical humor. Of course this is actually impossible, at least excepting a paranormal state of consciousness

[251] W.Bulloch, *The History of Bacteriology*, New York: Dover, 1979 (first ed. 1938). Bulloch gives much attention to Bassi's contribution (pp. 159-161, and p. 351); quotation on p. 161, my italics. S.Calandruccio wrote in 1892, G.C.Riquier in 1924, G.B.Grassi in 1925.

[252] The precursor fallacy (or *precursoritis*) has not been only Italian. After a documented analysis of Bassi's contribution, Caullery concludes with a typical note of merit: "Ces divers faits, longtemps méconus par les historiens de la biologie, montrent que A.Bassi a été le précurseur de la bactériologie"; see: M.Caullery, "Pasteur et la microbiologie," in: R.Taton (ed.), *Histoire générale des sciences. Le XIXe siécle*, cit., pp. 443-450, on p. 447.

[253] C.Castellani, "Bassi, Agostino," ST, s.v.

[254] G.Penso, *La conquista del mondo invisibile. Parassiti e microbi nella storia della civiltà*, Milano: Feltrinelli, 1973, p. 279.

[255] J.Rostand, *Lazzaro Spallanzani e le origini della biologia sperimentale*, cit., p. 187.

[256] H.A.Lechevalier, M.Solotorovsky, *Three Centuries of Microbiology*, New York: Dover, 1974, p.44.

[257] W.Bulloch, *The History of Bacteriology*, cit., p. 159.

[258] C.Castellani, "Bassi, Agostino," cit.

[259] A.Bassi, *Del mal del segno, calcinaccio o moscardino, malattia che affligge i bachi da seta e sul modo di liberarne le bigattaje anche le più infestate*, Lodi: Orcesi, part I, 1835; part II, 1836. A second enlarged edition was published in Milan in 1837.

[260] Bibliography in W.Bulloch, *The History of Bacteriology*,cit., pp. 313, 316.

[261] A.Bassi, *De la Muscardine*, Paris, 1836, quoted in: M.Caullery, "Pasteur et la microbiologie," cit., p. 446.

[262] Ib., pp. 164-165. Hanle was later one of Koch's teachers at Göttingen.

[263] C.Vittadini, "Della natura del calcino o mal del segno," *Mem. I.R.Istituto Lombardo*, **3**, pp. 447-512 (1852). Vittadini is defined a *precursore* in: M. Alippi Cappelletti, "La biologia italiana dell'Ottocento," cit., on p. 501.

[264] Morozzo's bibliography is in: *Il primo secolo della R.Accademia delle Scienze di Torino, 1783-1883*, cit., pp. 107-109.

[265] I.Guareschi, "La chimica in Italia dal 1750 al 1800," in *Suppl. Ann. Enc. Chim.*, **28**, pp. 393-470 (1911-12), on p. 405; A.Gaudiano, "Brugnatelli, Luigi Valentino," ST, s.v.; A.Gigli Berzolari, *Alessandro Volta e la cultura scientifica e tecnologica tra '700 e '800*, cit., p. 151, 156.

[266] G.Testi, *Storia della chimica, con particolare riguardo all'opera degli italiani*, Roma: Mediterranea, 1940, p. 235.

[267] P.Redondi, "Cultura e scienza dall'illuminismo al positivismo," cit., pp. 721-729.

[268] I.Cantù, L'Italia scientifica contemporanea, cit., s.v. "G.A.Majocchi".

[269] P.Redondi, "Cultura e scienza dall'illuminismo al positivismo," cit., p. 737.

[270] Ib., p. 724

[271] Quoted from: G.Testi, *Storia della chimica*, cit., pp. 236-237.

[272] Bizio's biblography is in: G.Provenzal, *Profili bio-bibliografici di chimici italiani*, Roma, s.d. (but 1938), pp. 127-132; the bibliography lists 106 titles.

[273] A.Mottana, "Lo sviluppo delle scienze della terra in Italia nel periodo 1840-1880," in: S.Di Sieno (ed.), *Scienze in Italia, 1840-1880*, cit., on p. 51.

[274] A.Gaudiano, "Piria, Raffaele," ST, s.v.

[275] B.J.Reeves, "Le tradizioni di ricerca fisica in Italia nel tardo diciannovesimo secolo," in: V.Ancarani (ed.), *La scienza accademica nell'Italia post-unitaria*, cit., pp. 53-95, on p. 57.

[276] P.Antoniotti, L.Cerruti, M.Rei, "I chimici italiani nel contesto europeo, 1870-1900," in: : V.Ancarani (ed.), *La scienza accademica nell'Italia post-unitaria*, cit., pp. 112-190, on p. 151.

[277] Meneghini introduced the fossil botany in Italy; see: M. Alippi Cappelletti, "La biologia italiana dell' Ottocento," cit., p. 500.

[278] At least this seems to be the opinion of Ugo Schiff, in a manuscript annotation in the first bound volume of *Gazzetta Chimica Italiana*, kept in the chemical library of Florence University. There is an excellent paper on the foundation of this journal: L.Paoloni, G.Paoloni, "La fondazione della «*Gazzetta Chimica Italiana*» (1870-1871)," in: A.Ballio, L.Paoloni (eds.), *Scritti di storia della scienza in onore di Giovanni Battista Marini-Bettòlo*, Roma: Accademia dei XL, 1990, pp. 244-280; Schiff's annotation is reproduced on p. 266.

[279] P.Antoniotti, L.Cerruti, M.Rei, "I chimici italiani nel contesto europeo, 1870-1900," cit., pp. 153-158.

[280] Società Italiana di Fisica, *Indice del Cimento e del Nuovo Cimento*, Pisa: Pieraccini, 1903.

[281] I.Guareschi, "La chimica in Italia dal 1750 al 1800," in *Suppl. Ann. Enc. Chim.*, **28**, pp. 393-470 (1911-12).

[282] Guareschi partecipated as volunteer in the war against Austria in 1866; v.: L.Cerruti, "Guareschi, Icilio," DBI, s.v., in press.

[283] Società Italiana per il Progresso delle Scienze, *Un secolo di progresso scientifico italiano (1839-1939)*, Roma: SIPS, volumes I-VI, 1939; volume VII, 1940.

[284] P.Redondi, "Cultura e scienza dall'illuminismo al positivismo," cit., p. 681.

[285] G.Cavallo, A.Messina, "Caratteri, ambienti e sviluppo dell'indagine fisica nel Novecento e la politica della ricerca," in: G.Micheli (ed.), *Scienza e tecnica nella cultura e nella società dal Rinascimento a oggi*, cit., pp. 1109-1162, on p. 1114.

[286] I used this type of information many times in: L.Cerruti, "Chimica e chimici in Italia. 1820-1970," in: C.Maccagni, P.Freguglia (eds), *La storia delle scienze*, cit., pp. 411-440.

[287] H.Kragh, *An Introduction to the Historiography of Science*, Cambridge: Cambridge UP, 1989, p. 111.

[288] P.Redondi, "Cultura e scienza dall'illuminismo al positivismo," cit., p. 685.

[289] This collective work represented a great effort by the learned community to present itself to a larger audience (e.g. to secondary school teachers) and to the Regime. It awaits a more attentive analysis.

[290] G.Micheli, "La cultura italiana di fronte alla scienza," in: G.Micheli (ed.), *Scienza e tecnica nella cultura e nella società dal Rinascimento a oggi*, cit., pp. XV-XXX, on pp. XXVI-XXVII.

[291] C.Maccagni, P.Freguglia (eds.), *Storia sociale e culturale dell'Italia*, vol. V, *La cultura filosofica e scientifica*, t. II, *La storia delle scienze*, Busto Arsizio: Bramante, 1989. It is not possible to use this book as a quick reference because it has no index of authors.

[292] V.Ancarani (ed.), *La scienza accademica nell'Italia post-unitaria. Discipline scientifiche e ricerca universitaria*, Milano: Angeli, 1989 (282 pages).

[293] F.Barbano, "Introduzione," in: V.Ancarani (ed.), *La scienza accademica nell'Italia post-unitaria*, cit., pp. VII-XXIX, on p. XXIII; V.Ancarani, "Università e ricerca nell'Italia post-unitaria. Saggio introduttivo," ib., pp. 1-36, on pp. 6-10.

[294] P.Antoniotti, L.Cerruti, M.Rei, "I chimici italiani nel contesto europeo, 1870-1900," in: V.Ancarani (ed.), *La scienza accademica nell'Italia post-unitaria*, cit., pp. 112-190, on pp. 115-149. The length of the analysis concerned the editor, who in the end permitted this vagrancy around the central theme.

[295] Ib., pp. 158-160, 183-186.

[296] The criteria depend critically on the social structure that we define as 'science'. The structure may be linguistic, epistemic, cognitive, institutional, and so on.

[297] The interested groups operate under the abbreviations PRISTEM (Project for Historical and Methodological Research) and GNFSC (National Group of Fundaments and History of Chemistry)

[298] L.Cerruti, "Scienze in Italia 1840-1880: una storia da fare," *Lettera Pristem*, n. 5, p. 29 (1992).

[299] L.Cerruti, "Mondi: corpuscolari e non. Fisica e chimica a confronto, 1840-1880," in: S.Di Sieno (ed.), *Scienze in Italia. Parte II*, cit., pp. 1-41. This tentative of using a larger context almost aborted because I worked on the epistemological and linguistic level, while most of the other historians worked on the institutional level. We also have to face the problem of commensurability between the different analytic frames of the historical research.

[300] S.Di Sieno (ed.), *Scienze in Italia, 1840-1880. Una storia da fare. Parte I*, Quaderni PRISTEM, n 4, 1993 (119 pages); *Parte II*, Quaderni PRISTEM, n 5, 1994 (156 pages).

[301] A.Asor Rosa, "La cultura," in: R.Romano, C.Vivanti (eds.), *Storia d'Italia*, vol. 4, *Dall'Unità a oggi*, t.2, Torino: Einaudi, 1975, pp. 821-1664.

[302] G.Micheli, "La cultura italiana di fronte alla scienza," cit., p. XXVII.

[303] Paolo Casini, "I silenzi di Clio," in: P.Nastasi (ed.), *Il Meridione e le Scienze*, cit., pp. 15-26, on pp. 16-17.

[304] R.Maiocchi, in a discussion during the workshop *Il fascismo e le scienze*, Rome, 10-11 October 1996.

[305] P.Redondi, "Cultura e scienza dall'illuminismo al positivismo," cit., p. 683.

[306] Quoted in: L.Besana, "Il concetto e l'ufficio della scienza nella scuola," in: G.Micheli (ed.), *Scienza e tecnica nella cultura e nella società dal Rinascimento a oggi*, cit., pp. 1165-1284, on p. 1187.

[307] G.Micheli, "Lazzaro Spallanzani e il pensiero scientifico del Settecento," in: G.Cherubini et al. (eds.), *Storia della società italiana*, vol. 12, *Il secolo dei lumi e delle riforme*, Milano: Teti, pp. 319-344.

[308] O.F.Mossotti, *Lezioni elementari di fisica-matematica date all'Università di Corfù nell'anno 1840-43*, Firenze, 1843, quoted in: L.Besana, "Il concetto e l'ufficio della scienza nella scuola," cit., p. 1204. Besana points out an analogous cultural frailty in the Italian positivists like Roberto Ardigò (1828-1920).

[309] L.Cerruti, *Temi di ricerca della chimica classica, 1820-1970*, Milano: Eurobase, 1990.

[310] L.Besana, "Il concetto e l'ufficio della scienza nella scuola," cit., p. 1197.

[311] V.Ancarani, "Università e ricerca nell'Italia post-unitaria. Saggio introduttivo," cit., p. 8-10; M.Rossi, *Università e società in Italia alla fine dell'800*, Firenze: Nuova Italia, 1976, pp. 1-4.

[312] M.Roggero, "Professori e studenti nelle Università tra crisi e riforme," in: C.Vivanti (ed.), *Intellettuali e potere*, Torino: Einaudi, pp. 1037-1081, on pp. 1070-1079.

[313] A.Gentili, "Paolo Frisi barnabita," in: G.Barbarisi (ed.), Ideologia e scienza nell'opera di Paolo Frisi (1728-1784), cit., vol. II, pp. 7-30, on p. 25.

[314] E.Codignola, "Gesuita moderno (Il)," in: *Dizionario letterario delle opere e dei personaggi*, vol. III, Milano: Bompiani, 1963, p. 594.

[315] Quoted in N.Badaloni, "La cultura," in: R.Romano, C.Vivanti (eds.), Dal primo settecento all'Unità, cit., pp. 697-984, on p. 972.

[316] G.Quazza, *L'utopia di Quintino Sella. La politica della scienza*, Torino: Istituto per la Storia del Risorgimento Italiano, 1992, p.93.

[317] Ib., p. 99.

[318] Ib., p. 106.

[319] Quoted in: E.Passerin d'Entreves, "Ideologie del Risorgimento," cit., p. 323. Tommaseo's condemnation of the religious intolerance recall the unbearable conditions of Jews living in the ghetto in Rome. On 5 July 1827 Pope Leon XII had again put the Jew community under the heavy control of an edit of Pope Pio VI (5 April 1775, Editto sopra gli Ebrei). On the Jew side this edit has been judged "a monstrous codex of denial of whatever human dignity"; A.Milano, Storia degli ebrei in Italia, Torino: Einaudi, 1963, p. 296. The official title of the edict was: Fra le pastorali sollecitudini [Amid pastoral care]; it was one of the first official acts of the Pope, elected on 15 February 1775. The hard treatment reserved to Roman Jews has been correlated to Pio VI's attitude entirely adverse to enlightenment; see: M.Rosa, "La Santa Sede e gli ebrei nel settecento," in: C.Vivanti, Gli ebrei in Italia, vol. II, Dall'emancipazione ad oggi, Torino: 1988, pp. 1067-1087, on p. 1084; D.Menozzi, "Tra riforma e restaurazione. Dalla crisi della società cristiana al mito della cristianità medioevale (1758-1848)," in: G.Chittolini, G.Miccoli, La Chiesa e il potere politico dal Medio Evo all'età contemporanea, Torino: Einaudi, 1986, pp. 767-806, on p. 780.

[320] In the Italian context 'democratic' was the appellative of politicians of a moderate left, mostly republicans like Mazzini and Cattaneo.

[321] G.Montanelli, *Introduzione ed alcuni appunti storici sulla rivoluzione d'Italia*, Torino, 1851; quoted in: A.Desideri, Storia e storiografia, Firenze: D'Anna, 1987, p. 654.

[322] G.Quazza, *L'utopia di Quintino Sella*, cit., pp. 139 and 485.

[323] A judgement by Quazza; ib., p. 497.

[324] Ib., pp. 497-498. The Italian text is: *Non abbia paura scienza progresso - che paura io credo abbia dettate disposizioni fatte prevalere da fazione gesuitica contro convinzioni prelati più illuminati.*

[325] L.Besana, "Il concetto e l'ufficio della scienza nella scuola," cit., pp. 1186-1189. Besana discusses also a long list of Jesuits' contributions in physical sciences and mathematics, published in 1941.

[326] G.Galasso, *L'Italia come problema storiografico*, Torino: UTET, 1979, pp. 72-73.

[327] Ib., p. 101.

[328] Ib., p. 102; the words are by Graziadio Isaia Ascoli, a great linguist, and referred, originally, to Germany.

[329] A.Gramsci, *Il Risorgimento*, cit., p.62.

[330] For example, when commissions visited the scientific and public health institutions of one State, the reports were published and circulated in other States; see: *Rapporti della commissione incaricata dalla sezione di medicina del settimo Congresso degli Scienziati italiani di visitare gli spedali civili e militari di Napoli e Aversa*, Firenze: Società Tipografica, 1845.

[331] L.Cerruti, "Procedure conoscitive e culture disciplinari. Un'analisi storiografica," in: G.Battimelli, E.Gagliasso (eds.), *Le Comunità Scientifiche tra storia e sociologia della scienza*, Quaderni della Rivista di Storia della Scienza, n. 2, 1992, pp. 83-122. In this book, Battimelli and Gagliasso edited the lectures and the communications given in a workshop on scientific communities (Rome, 18-20 April 1991).

[332] M.Polanyi, *Personal Knowledge. Towards a Post-Critical Philosophy*, London: Routledge, 1983 (1958).

[333] L.Cerruti, "Un profilo di Lothar Meyer. I. I lunghi anni della formazione," *Chim.Ind.*, 78, 863-868 (1996).

[334] C.Pighetti, *Carlo Matteucci e il Risorgimento scientifico*, cit., p. 16.

ARNE HESSENBRUCH

THE SPREAD OF PRECISION MEASUREMENT IN SCANDINAVIA 1660-1800

INTRODUCTION

The history of science in Scandinavia has usually been written on the basis of the published sources of institutions such as the Academies of Science, or of the monographs by Professors of the Universities. If this were all the scientific activity, it would have been a veneer, an activity limited to an urban elite, very remote from the experience of the vast majority of the population. Such a description of science might possibly be appropriate for the St. Petersburg Academy with its imported Professors, which remained separate from the lives of most Russians and failed to have the modernising impact upon Russian society which its benefactor, Peter the Great, had hoped.

But in Scandinavia, much of the scientific activity was closely related to everyday concerns in agriculture and the levying of state revenues. This role of Scandinavian science, with its tentacles reaching out to the furthest corners of the lands, has hitherto been ignored, primarily because it requires much more work than simply consulting the conveniently bound transactions of the Academies; but also because the lowly forms of measurement performed at every town gate or at every port has not been recognised as science.

However, measurement in general and precision measurement in particular is rightly seen as a scientific activity, and as I will show, the precision of measurement was at issue wherever merchandise was weighed or evaluated. Moreover, much of the activity in the Academies was aimed at providing the means for the state to establish a metrological regime, a network of calibration, so that measurement everywhere was as precise and reliable as possible.

There are some studies which have addressed the avalanche of numbers. First of all, *The Quantifying Spirit*, much of which is based on Swedish material, maps the growth of numbers in the 18th century.[1] However, more recently, the focus has shifted towards the workings of the state and its requirements of scientific activity. Norton Wise's *Values of Precision* addresses the questions of the bureaucratic state, the problems of controlling territory, especially at a distance, standardization of measurement as a means of controlling measurement at a distance, and local dissent in the face of metropolitan decrees.[2] These issues are just the ones I will address in this essay.

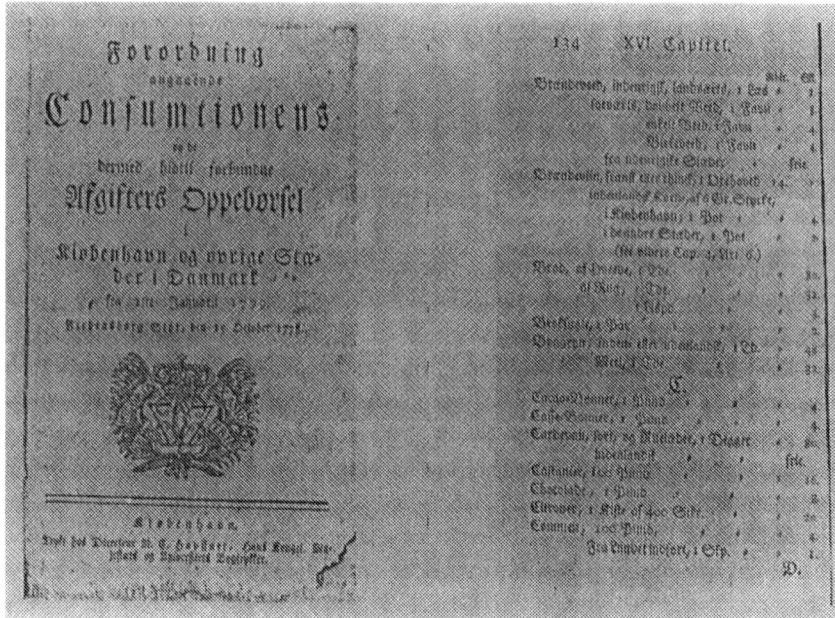

Illustration 1.

But I want to bring attention to the fact that numbers are not always compatible, in the sense that two apples plus two oranges does not yield four, unless you change the unit to pieces of fruit. You can not perform common arithmetic unless the units are clearly defined. This is the case for the measurement performed in Scandinavia in the period concerned. Customs or excise tariffs list commodities *in particular units* along with a fee, in money. The Danish excise for 1779 (Illustration 1: Forordning angaaende Consumtionens...'), for instance, lists the fee for barley groats (Byggryn) as 48 shillings per barrel, and for French or Rhinelandish spirits (Brændeviin) as 14 Rixdaler for a hogshead and 4 shillings per 'pot' for the domestic variety when sold in Copenhagen (2 in the other towns). Performing calculations and comparing amounts of commodity was difficult with the wealth of units in use - both in the sense that each commodity was measured in its own unit, and in the sense that units varied from place to place. A set of four tables on weights and measures illustrates the need for a conversion between units (Illustration 2).[3] Under 'Dansk Vegt' (Danish weights) we see both the units mentioned above: 1 Skippund

THE SPREAD OF PRECISION MEASUREMENT IN SCANDINAVIA 1660-1800   181

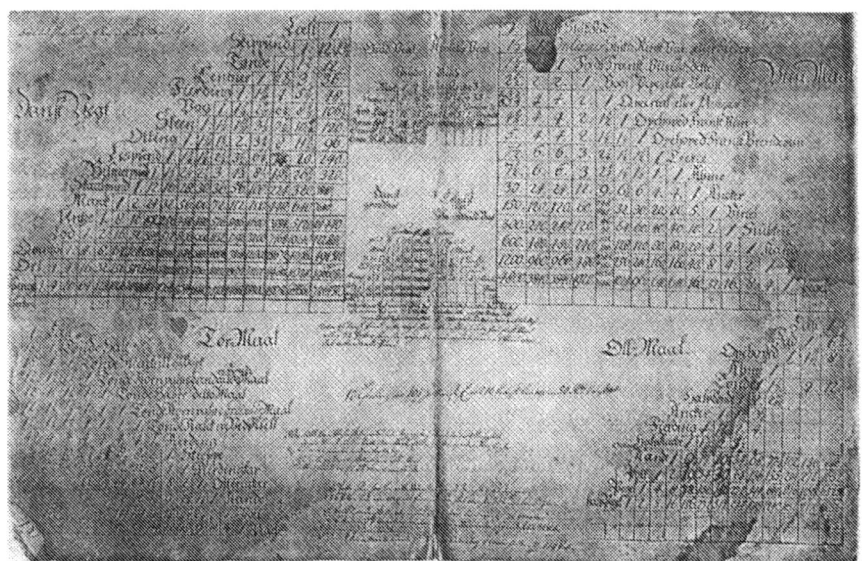

Illustration 2.

= $1^3/_7$ Tønde. The other tables refer to 'Viin Maal' (wine measures), 'Øll-Maal' (Beer measures), and finally 'Tør-Maal' (dry measures) which includes 'Tønde, Kornvahre Landgilde og Tiende Maal' (grain barrel for manorial dues and tithes) and also a grain barrel for ordinary measures.

The state also spawned increased quantification in the measure of land and here the units were of length and surface area. The measurements took place with rods of standard length. With this and an assessment of the fertility of the land, the likely yield of the plot was determined, in barrels of grain. According to this value, the tax, was set. The commodity land[4], just as the commodity spirit, was measured in a unit which was then related to money. A paper published in the transactions of the Copenhagen Academy of Sciences was explicit about the relation between land value and value in cash, as the title itself reveals: 'How to determine the size of land which can be used for payment of the tithe'[5] (the tithe, as we have seen in the last paragraph was usually paid in barrels of grain).

The period from 1660 to 1800 is not just a period of quantification, in which more measurements are carried out and more numbers created and used in administrational matters, it is also a period where state administrations cry out for order in the use of units. This was never straightforward. Both Scandinavian states decreed standard weights and measures in the late 17th century, but the practice on the ground ignored the decrees to a very large degree. As we shall see, imposing a standard required educated civil servants who would police the use of weights and measures, backed up by severe penal measures.

This paper will illustrate the explosive growth of measurement in Scandinavia by delineating the gradual introduction of standard weights and measures while attending both to the role which elite science played in this development and also following, as far as is possible, the activities on the ground. In the following I will start by outlining the development of Scandinavian elite science (Section 1), and refer to that part of their work which was related to the problems of measurement which interested the state (Section 2). The main issue is metrology, the introduction of standard weights and measures which were promulgated from the metropoles and intended to eliminate all local diversity. Both Scandinavian states, Sweden-Finland and Denmark-Norway, employed precision measurement in the surveying of land at first with a view to rationalising and making more efficient the business of raising land tax, and later also with a view to promote enclosures which increased the yield of the land. Standard length was one contribution to the efficient management of changing land property regimes. Both states increasingly turned to customs and excise which involved measurement of barrels, cartloads, and bottle contents. Here too, standards of weights and measures took the wind out of disputes in matters of taxation. All this employment of precision measurement and standard units in fiscal matters is the topic of Section 4. Before entering into that field, I thought it helpful to delineate the requirements of the two Scandinavian states for increasing revenue. This is given in Section 3, which is the only section that could be described as general history, and not by itself a part of history of science. However, my argument is that the growth of precision measurement and metrology is a response to a fiscal demand, and therefore the realities of the fiscal-military states ought to be presented. In the fifth section, I will describe in more detail three aspects of the management of Scandinavian metrological regimes. One aspect is the professionalisation of the people on the ground doing the measurements, especially of the surveyors. Another is the production of scientific instruments and their adaptation to the reality of raising taxes. A last issue is the unit of value employed for reckoning: money. The land is evaluated in terms of money, and customs regulations list fees for a particular amount of each particular commodity. There is always a commodity, measured in length, surface, volume, or weight, and related to a sum of money. But money never yields stable value. Coins themselves were not trusted as they were minted as substandards, clipped or otherwise debased. Money was thus subject to the same kind of management as other metrological regimes and can not be taken for granted. Finally, I will discuss questions of morality and authority. Cheating and measuring were intimately related in contemporary minds, and standard weights and measures explicitly addressed the question of fairness. The state intervenes in exchange relations by providing authoritative measurement, purportedly making fraud more difficult - but certainly also making tax fraud more difficult. At any rate, Scandinavian societies were profoundly changed by this state intervention at a local level, in a sense rendered more rational.

1. THE VENEER OF ELITE SCIENCE

In 1660, Scandinavia was dominated by agriculture. The vast part of the population lived in rural villages, and the market towns were small. People on the land lived in what was basically a subsistence economy, with trade in agricultural goods restricted to surplus production. In the dual kingdom of Denmark-Norway, the surplus in Denmark was exported to Norway. Within the Swedish empire, much grain was imported into Sweden proper from the colonies on the southern Baltic coast. Sweden also exported timber, tar, iron and copper to the trade centres in the Netherlands and England. Fish and timber was also exported from Norway, and the mine in Kongsberg produced metal for coinage in the mints of Denmark-Norway. There was thus a small but significant amount of trade.[6]

By 1800, Scandinavia had been integrated much further into world trade and while agriculture was still by far the most important economic sector, the infrastructure had changed, the population of towns and especially of the capitals Copenhagen and Stockholm had grown considerably. The property structure had changed towards the ownership of land through several waves of enclosure, and both barter and payment of taxes in kind had been largely replaced by exchange involving money.[7]

The period between 1660 and 1800 also saw the development of scientific institutions modelled on those found in London, Paris and other European centres. Prior to 1660, there had existed universities with the usual medieval structures focussing on theology, law and medicine. The Universities of Uppsala and of Copenhagen had been founded by papal bulls in 1477 and 1478 respectively.[8] Åbo University in present day Turku, Finland, was founded in 1640.[9] Lund University was founded at the beginning of the period concerned, in 1666.

But the earliest Scandinavian institutions which explicitly pursued science were founded in the 18th century. At Uppsala University in 1710, a Collegium curiosorum was founded, which was soon renamed Royal Society of Science at Uppsala (Kungliga Vetenskapssocieteten i Uppsala). In the 1720s the *Acta literaria Suecia* was published under its auspices. The Royal Swedish Academy of Sciences was founded in 1739. This institution was not a university. Just as with the Parisian Académie des Sciences and the Royal Society of London, it heralded a new infrastructure for scientific discourse. The differences between Uppsala University and the Stockholm Academy has been described very graphically in terms of a number of dichotomies: provincial university - capital; Latin - Swedish; archival studies - utilitarianism.[10]

In the 18th century and to this day, the Scandinavian neighbours have kept an eye on each others' institutional innovations. When the Stockholm Academy of Sciences was founded, the Western Scandinavian power had to have a similar institution. The Copenhagen Royal Academy of Sciences (Det Kongelige Danske Videnskabernes Selskab) was founded in 1742, three years after the Stockholm Academy.[11] Throughout the 18th century a small number of similar scientific societies was founded, including The Learned Society of Trondheim (Det trondhjemske lærde Selskab) of 1760, which in 1767 became the Royal Norwegian Society for the Sciences (Det Kongelige Norske Videnskabernes Selskab), the Royal Danish Society of Agriculture (Det kongelige danske landhusholdningsselskab) of Copenhagen (1769) modelled on the constitution of the London Royal Society of Arts, and the Royal Academy of Science and Letters of Gothenburg (Kungl. Vetenskaps- och Vitterhets-Samhället i Göteborg), founded in 1778.

The Stockholm Academy was reasonably successful in the first half-century of its existence, much more so than its Copenhagen counterpart. This can be seen clearly in the fate of the transactions of the Academies, both of which were in the national language, which obviously caused some problems with the distribution beyond the national boundaries. The Swedish Academy was still quite successful. It managed to arrange exchange relations with the Royal Society (*Handlingar* against the *Phil. Trans.*) before the Royal Society realised that the Swedish journal was not in Latin. Of course, both the Berlin and the St. Petersburg Academies employed Latin for their publications. However, both the Stockholm and the Copenhagen Acedemies were involved in the standardisation of the national language, again as a part of the construction of the nation state. The Transactions of the Danish Academy were not at all in demand. By 1815 it was decided to publish abstracts in a foreign language to at least have some international distribution. One reason for the Swedish success and the Danish failure was that the Stockholm Academy was granted the monopoly for the publication of almanacs, the sale of which subsequently provided its main income.[12] The Copenhagen Academy failed to do so.

Not only were the institutions modelled on those of the big European powers, most of the Scandinavian individuals travelled for long periods in the urban centres in order to pick up information or techniques which could be employed back home.

In order to illustrate the international contacts provided through such travels, a brief presentation of some half dozen Swedes and then Danes will be given. This is not intended as a comprehensive list; it should just give a flavour of the Europewide network spanned by members of the two Scandinavian academies.

First the Swedes: Emanuel Swedenborg (1688-1772) spent three years in Oxford and London and two in Holland and Paris. While abroad, he sent books and instruments to friends in Uppsala. Carl Linné (1707-1778) who has been credited with the first scientific botanic taxonomy, received a doctorate from Harderwijk University in Holland, and spent time in Oxford, Leiden, and Paris, meeting amongst others Jussieu. Anders Celsius (1701-1744), of thermometer scale fame, travelled to Berlin, Nuremberg, Italy, Paris (where he met Maupertuis) and London where he purchased scientific instruments. Mårten Triewald (1691-1747), one of the co-founders of the Stockholm Academy, spent two years as a trader in Königsberg and 10 years in Britain as scientific lecturer and steam engine consultant. He also collaborated with the Board of Mines in importing steam engines to Sweden.[13] Samuel Klingenstierna (1698-1765), an astronomer, spent four years travelling through Germany, Switzerland (where he met Bernoulli), France (where he met Clairault and Fontenelle), and England. Jacob Faggot, member of the Royal Academy of Sciences and soon to become head of the Board of Surveyors, published *The obstacles to Swedish agriculture and how to overcome them* in 1746. He was a major force behind the enclosures movement and features prominently in this story. Henrik Kalmeter (1693-1750), the Auscultator in the Board of Mines, visited a great number of English mines in 1719-1720.[14]

The Danes: Ole Rømer (1644-1710) is now known as the first person ever to measure the speed of light. He became a member of the Académie des Sciences and he lived in Paris from 1672 to 1681. He is the only person mentioned here, who was alive at the beginning of the period covered here. This is no coincidence: it is only with the institutions set up in the 18th century that individuals engaged in waht we recognise as science can be found in any number. Jens Kraft (1720-1765) published on the theory of machines, on monads, and on Newton's theories. He spent a total of two years in Marburg and Basel. Christian Gottlieb Kratzenstein (1723-1795), originally a Saxon, who was a member at the St. Petersburg Academy before being called to the University of Copenhagen. He wrote monographs on medical electricity and on experimental physics. Christian Friis Rottbøll (1727-1797) studied with Linné in Uppsala and spent 1757-1761 in Holland and France. He became Professor of botany at Copenhagen University. Peter Christian Abildgaard (1740-1801), spent four years at the School of Veterinary Studies at Lyons. He wrote a string of monographs on how to improve livestock breeding. Thomas Bugge (1740-1815) is known to historians of science because his diary of meetings with French savants (such as Lalande, Laplace, and Legendre) and his inspection of scientific and educational establishments during the 1800 Paris conference promulgating the meter in 1800 has been republished.[15]

These lists reveal that the Academies' members had considerable international experience and contacts. But even so, we are only talking about some dozens of individuals engaged in an intellectual discourse which was remote from the experience of the peasants. The science of the Academies was a metropolitan, polite affair, which general historians mostly disregard because it does not seem to have had any impact on the development of the country as a whole. For example, the import of a steam engine to the Swedish copper mining district which Mårten Triewald oversaw was a financial failure and almost all the attempts to use new technologies supported by the Copenhagen government to foster new manufactures failed.[16] Agriculture remained by far the most important economic sector.

2. ELITE PUBLICATIONS ON METROLOGY

Like a tip of the iceberg, the scientific work reveals the preoccupation of the state with measurement, especially of land and merchandise, the most important of which was grain.

Already Ole Rømer was much more engaged with the Danish State than his measurement of the speed of light might lead us to believe. He was put in charge of the reorganisation of weights and measures and of the matriculation, a surveying and evaluation of land for the purposes of establishing taxes and simplifying the task of raising revenue. He ended his career as Chief of Police.[17] Rømer is no exception to the rule. Metrological work amounts to a substantial amount of the publications of the Academies.

For example, in 1740, Anders Celsius reported on comparisons which he had made between the standard rod kept in the Stockholm town hall and various European standards, including those of Rome, Paris, London, and Leyden.[18] Jacob Faggot published an article on the problem for standards caused by the expansion with temperature, and one on the use of the decimal system with standards.[19] There were attempts to base a standard rod on the second pendulum,[20] and no less than two new members gave their inaugural paper on weights and measures.[21] The question of the evaluation of the quality of land was discussed.[22] The role of surveying also included measurements of the highest precision for military purposes and the tasks of the Board of Surveyors was split between the military projects and the more mundane (and less high precision) measurements for civil purposes[23].

In high science, weighing was very important. As Anders Lundgreen has argued, the chemical revolution is related to the increased employment of the chemical balance for mining and other commercial purposes.[24] The transactions of the Academy of Sciences also bears witness to the importance of the balance. (Some of the already mentioned publications on the determination of volume dealt with weights too.) Anders Berch, Faggot, and E. Runeberg were the main contributors.[25] The transactions of the Royal Academy reveals the widespread academic concern with volume measurement. Anders Berch who published several comparisons of weights also published on jugs.[26] The Academy member and instrument maker Daniel Ekström published on a grain tester which established the quality from a measurement of specific weight (I will discuss this instrument below), a topic which also concerned E. Runeberg.[27] Testing of alcohol content in conjunction with volume measurement was employed to evaluate beer, wine and spirits.[28] The geometry required to work out volume from the exterior measures of vessels was also studied.[29]

Many of the problems of quantification for the Danish state are also reflected in the publications of the Copenhagen Academy of Sciences. The first publications were almanacs which included tables of numerous countries' measure of length; lists of land tax by Danish provinces (1760); and tables on foreign measures of volume (1761-1763). Other publications dealt with Danish measures of length[30] (it was important to determine length for the precise evaluation of postal fees); new calculations of land tax liability[31], the mathematical apparatus for working out the plot of land that could replace barrels of grain as a tithe payment[32], geometric and economic knowledge of enclosure[33], weights and measures in coinage[34], assaying of silver and gold[35], and the evaluation of alcoholic spirits[36].

3. SITUATING METROLOGY WITHIN THE DEMANDS OF THE SCANDINAVIAN STATES

This substantial volume of metrological work by the scientific elite can only be understood when situated within the demands of the state. This section sets out to contextualise these demands. The historical context reveals the need for ever-widening tentacles of state-guided and state-controlled measurement.

It is convenient to divide the period covered here up into two parts, 1660-1720 and 1720-1800, the first period characterised by the introduction of military-fiscal states with state machineries reorganised to raise sufficient taxes for the purpose of waging war, to a large extent with the purpose of gaining maximum profit from Baltic trade. The second period is characterised by a further increase in the cost of militarisation but both Scandinavian powers lost the means of controlling the Baltic, especially faced with the growing military might of Russia. This led to the need for an increase of efficiency within the heartlands, both with regard to the state machinery levying taxes, and with regard to the introduction of new systems of agriculture, involving private property, enclosure, and investment in improvements of draining, machinery etc.

3.1 1660-1720

In the period from 1660 to 1720, both Scandinavian states were very much aggressive expansionist military states, constantly interlocked in wars. In the 17th century Sweden had gained the upper hand, by 1660 controlling most of the Baltic coastal areas. In the 1690s, more than a fifth of Sweden's revenue came from the Baltic provinces, such as Ingria which supplied Sweden proper with large quantities of grain, in 1696 (a year of famine) as much as 800,000 tons.[37] Denmark-Norway, having lost most of present-day southern Sweden sought every opportunity for revenge. In the Great Nothern War (1709-1720), Sweden was faced not only with its traditional Scandinavian rival but also the expanding Russia of Peter the Great and with Poland and Prussia. The Swedish empire crumbled in the face of this massed opposition.[38]

In this period a new kind of state developed throughout most parts of Europe, and Scandinavia was no exception. New expensive war technologies were developed and the new type of power corporation which came into being was required to organise and harness the new powers. The state thereby took over some of the traditional duties of the nobility in that the monopoly of military violence shifted towards the state.

As a result of this development, the financial requirements increased considerably. The growth of the fiscal state therefore went hand in hand with that of the military state. This was the case in France under Louis XIII, and also in Denmark-Norway and Sweden-Finland. In all countries, the traditional source of revenue (income in kind from Crown land) was supplemented by and to some extent replaced with direct taxation. Concentration of power, bureaucratisation, and centralisation of the state administration ensued. Most European states developed away from the medieval structures based on the estates towards the new political structure with an emancipated state power. Absolutism was the result in Scandinavia too: it was introduced in Denmark in 1660 and in Sweden in 1680.[39] Within this new fiscal-military state, scientific instruments were to play an important role for the raising of taxes. They became an integral part of the regular state machinery.

During the 16th century the Swedish government had consisted of a number of centralized administrative bodies and a local administration consisting mainly of bailiffs. In the first third of the 17th, this organisation was expanded. Central government departments grew in number and now assumed the collegial form that they were long to retain (the colleges were similar to present-day ministries in that they were task-directed: fiscal, church, military, etc., whereas the previous organisation had been organised along other categories, for instance geographical[40]). To them was appended a local administration led by county governers. Modernisation and disciplining of the administration was also aimed for through the training of bureaucrats.

In 1634, a reform was pushed through parliament leaving no room for the personal rule of the king; instead an administrative juggernaut was created: a permanently organised central and local administration, in which the different bodies had clearly defined duties. At the head were the five high officers of state, each of them the head of an administrative board or college.

One of these colleges was the *kammarkollegium*, which was generally responsible for the state finances.[41] Within this college there were several departments which employed scientific instruments for the purpose of regulating and maximising the income of the state. The income of the state came mainly from the following sources: land tax, tolls levied on trade in the Baltic Sea, excise, and the sale of copper. All involved the use of scientific instruments and in all departments one can discern a development towards greater precision in measurement along with greater efficiency in levying. The land tax involved evaluation performed by a surveying board, set up in 1628 and soon answerable to the *kammarkollegium*. The tolls and the excise required primarily an estimation of volume but also weighing, while the copper trade first of all required weighing. In the Swedish language 'weights and measures' are divided into three (*mått, mål* and *vikt*) referring to the measures of length, volume and weight, corresponding to the main techniques of evaluation employed by the fiscal state.

In Denmark-Norway too, the absolutist state reorganised its structures on the basis of the collegiate principle. The nobility lost its monopoly on top administrational jobs and a measure of meritocracy was introduced. It was due to this new system that people like Ole Rømer could rise to important posts within the Danish administration.

The absolute monarchy ascending power in 1660 was lacking the most obvious instrument to form a national policy, which are so obvious today: statistics. Exact knowledge about area, land use, production units and population was not availabe. The necessity of leaving power at the local level to the landed gentry amounted to a lack of perspicuity for the central government. All the government could ask of the gentry was to deliver a certain amount of tax. How that tax was raised was left to the gentry.[42]

Denmark-Norway was in need of a new metrology to lower transaction costs and an efficient taxation system; preferably one that the peasantry perceived as a step forward leaving random aristocratic misrule behind, in the process justifying absolutism. A new registration of taxes to allocate the burden of revenue more evenly and justly had first priority, but since taxes were generally paid in kind the new weights and measures by themselves offered a facility. The gentry still held the local authority and was still responsiblity for collecting tax, in return for which they were exempted from their personal tax-burden. Since the register was given first priority it had to do without the metrology reform. Consequently it was founded on the old measures, not only the first register of 1664 which was precipitated, but also the second one which was begun in 1684, immediately following the Scanic War (1675-79) where the Danish navy defeated the Swedish one and the Danish army recaptured the three provinces, Scania, Hallandia and Blekinge, lost in 1659, only to loose them again - this time permanently. The Danish-Norwegian Treasury coffers were empty.

3.2 1720-1800

The period after 1720 only increased the fiscal pressures. Denmark-Norway essentially entered into no armed conflict until the Napoleonic Wars, but it still armed heavily faced both with Russia and the old enemy Sweden-Finland . The latter was engaged in constant warfare in a futile attempt to contain Russia. Russia's emergence as a Baltic power rendered control of Baltic trade increasingly difficult for the Scandinavian states to control Baltic trade and their grip on customs and tolls was broken[43]. Accordingly, both Scandinavian states had to turn their efforts towards increasing revenue and economic strength from within.

Sample annual Swedish state revenue 1722-1806 (in silver daler coins)[44]				
	Land tax	Indirect tax	Loans	Total revenue
1722	1,825,468	1,077,315		6,215,740
1734	1,958,124	1,390,300		5,604,684
1746	1,643,272	1,409,700		7,307,307
1758	2,146,042	1,660,000	8,010,000	21,374,645
1770	2,377,361	1,090,674	90,050	16,379,872
1782	780,344	549,482	3,652	3,190,655
1794	786,544	750,000	87,000	4,984,800
1806	844,882	807,651	33,333	4,097,848

Forms of income not mentioned include exports (copper, iron, tar, wood) and subsidies. The public sector borrowing requirement can be shown to coincide with the wars which Sweden fought: 1741-1743 War against Russia; 1757-1762 War against Prussia; 1788 War with Russia (and Denmark); 1805: War against Napoleon. The most serious public sector borrowing requirements fell within the years of the Prussian war.[45]

The same fiscal pressures apply to the Danish-Norwegian state; its debt forced ever higher by the mere threat of warfare[46]:

Total debt of Denmark-Norway (in million Rigsdalers):										
1700	1730	1750	1754	1760	1763	1765	1774	1784	1790	1799
1.5	3.2	1.4	1.1	8.1	17.6	20.1	25.2	36.0	46.7	47.1

Both states responded to the requirements of war with increased internal revenue. They became capable of levying ever higher taxes partly through refinement of the state machineries but also helped by a buoyant economy. The price of grain rose considerably towards the mid-18th century, and towards the end of the century, the new forms of agriculture were beginning to produce higher yields too. This boom led to an emergence of a private sector of more than negligible proportion. The availability of credit increased. In 1660, the only accumulation of capital had been in the hands of the king, and to a lesser extent in the church coffers. The landed nobility had received dues in kind and not in money. Land was not bought or sold to any significant degree: farmers tilled the land owned by the gentry against an annual manorial due. By the mid-18th century, banks were coming into existence, and exchange involving money was becoming increasingly common. Hitherto common and open land was to be parcelled out, enclosed and owned by the farmer tilling it. This momentous development relied very much on the land becoming a commodity for which the evaluation and pricing of land was a necessity.

Whereas the late 17th century had seen the revolutionary development of European states with a juggernaut administration dealing in a routine way with the collection of revenue, the 18th century required even higher efficiency and greater administrational sophistication. The late 17th-century Scandinavian states introduced new tax registers. The increasing privatisation of land and the increasing role of indirect taxation required a wider network of state control with measurement. This is true in the case of enclosures and the introduction of private ownership of land, where the quantification provided by surveyors is crucial. And it is equally true of customs and excise where the levying of tolls was made more efficient through the introduction of national standards of weights and measures. Quantification became more and more important for the functioning of the state, and the requisite administrational machinery for the control of measurement and for the standardisation of units increased in size and sophistication.

In the following section on the means of raising tax, I will outline the administrational machineries for the evaluation of tax, pointing to the measurement required, and the precision expected of that measurement.

4. SCIENCE IN THE SERVICE OF THE FISCAL SYSTEM

4.1 Direct taxation, 1660-1720

Before 1660, between a third and half the income of the Danish state came from Crown lands. Direct taxation accounted for half of its revenue from the early 1660s onwards. In addition, indirect taxes amounted to about a third. This corresponds roughly to the percentages of France under Colbert, a state notorious for its high taxation. In the ensuing years the military expenditure resulted in debts of such proportions that the Danish Crown was forced to sell land with the result that only a small percentage of revenue accrued from Crown property. The absolutist Danish state now had to rely on the efficiency of taxation to a much higher degree.

Swedish Crown land also varied in the second half of the 17th century. From 1655 onwards, land belonging to nobility was reclaimed by the Crown (the so-called *reduktion*). It was politically feasible because most high noblemen were simultaneously high civil servants, and they would have to accept severe salary cutbacks if they did not agree to the *reduktion*. The state used the reclaimed land for parcelling out to soldiers as a salary. This reduced the cash flow requirement of mobilizing and maintaining an army. The important consequence for our purposes is that the parcelling out increased the demand for surveying.

But the need for precise measurement of land was more acute when taxation was involved than when the state confiscated or donated land. The Swedish state was particularly keen to establish an uncontroversial tax basis in the southern provinces captured from Denmark in 1660. Their mapping and adaptation to the Swedish tax system was partly carried out under the leadership of former Danish aristocratic residents. By 1683 most of Sweden including Scania was mapped roughly.[47] Once again, the Copenhagen government decided to copy a Swedish institution. One of the Scanian nobles, Knud Thott, was induced to switch his allegiance back to Copenhagen where he obtained high positions in the central administration.[48] He brought administrational experience from Sweden with him, and the Danish tax-register was modelled on the Swedish cartographical project. In order to ensure that the required expertise was available on the ground, Swedish surveyors were hired. In this way, surveying and the generation of tax registers were pursued in a similar manner in both Sweden-Finland and in Denmark-Norway.

It is hardly surprising that the registers caused friction. In the Danish case, some local gentry were at first given the right to deviate to some extent from the results of the measurements. To address the discontent in a more thorough way, a second land register was carried out with more effort and with more attention to local specifics, giving the Copenhagen government an overview over the ability to pay tax, county by county, property by property. It was finished in 1688. This also reduced the importance of the landed gentry as a kind of middle management.

The Danish land register enabled the state to make the *hartkorn* tax the pillar of the whole taxation system, that is to say that tax was estimated in a particular number of barrels of specified grain: usually rye or barley. Note the nationwide promulgation of one unit only: the barrel of grain. Most other taxes were gradually absorbed into it, that is they were estimated and levied simultaneously. Military dues of most of the rural population were transformed in 1676 to a tax to be paid in money (one of the steps towards a money economy). The various military dues of the nobility followed three years later. In 1692, all such kinds of tax were formally absorbed into the *hartkorn* tax.[49]

Thus, both Scandinavian powers surveyed, mapped, and created tax registers for the purposes of taxation in the late 17th century. In the 18th century these endeavours intensified.

4.2a Direct taxation 1720-1810: Sweden

In 1725, most of Sweden-Finland's agriculture was run on the age-old principles of a village community sharing most of the soil, without rotation of crops. The rotation of seed and trees was used only in a few areas. In a very few areas, a rotation of more than two crops was employed. The age-old agricultural practices stood in the way of the more modern techniques which Jacob Faggot wanted to introduce. The situation was essentially the same in Denmark-Norway. In both countries enclosures and private ownership was eventually to increase the yield, but the obstacles of the time were significant: breaking up the social structures of the village and changing rural life radically. Small crofters in particular lost out. The initiatives for land reform came from above, and as we shall see, there was continued inertia and outright resistance to the new schemes.

Jacob Faggot was a most vociferous proponent of enclosures. As a part of his campaign he became Head of the Board of Surveyors. Faggot's argument was that allowing the individual peasant control over his land, excluding grazing animals and other peasants, enabled him to sow what he wanted, rotate crops, and generally work undisturbed to get the most out of the soil. According to Faggot, the old ways limited private initiative and thereby curbed improvements. In his initial scheme, consolidation (the exchange of disconnected plots with the aim of giving each owner one connected piece of land) was to be voluntary. The exchange of land was to be seen to be just by the measurements of the surveyor.

The voluntary scheme was largely unsuccessful with few peasants coming forward.[50] Consolidation was considered impossible in areas with great local diversity in the quality of the land. Faggot promoted enclosure vigorously for decades. Since at first the exchange of land was voluntary, his first strategy was to diminish the peasants' fear of the risks involved. In 1749 a new instruction for surveyors was introduced in which the administrational processing of protests was streamlined. Faggot wanted the measurements done for the purpose of tax evaluation to be used for consolidations too. More importantly, in 1752, the king agreed to instruct all prefects that consolidation had to be performed as long as just one peasant demanded it, regardless of the objections of fellow villagers. Faggot requested the unprecedented number of 72 surveyors to process consolidation, but in vain.

However, success ensued in 1756, when the *kammarkollegium* worked out a report concluding that consolidation was possible in the entire country. The year after, it was resolved to carry through a consolidation with 10 surveyors in all of Scania, the province which produced most of the country's grain. Here, the ennobled Scottish immigrant, Rutger Maclean, had introduced modern crop rotation cultivation. The consolidations prompted a lot of peasant resistance, as usual because of the destruction of traditional village life and because of disputes over the new distribution of land.

Towards the end of the 18th century, consolidation had developed into one of the most important political issues. Resistance to the reforms was as vocal as the disruptions of traditional life was thorough. A 'Patriotic Economic Society' was founded arguing the advantages of the new agricultural system. The most modern agricultural techniques, however, were only introduced at large manors, the owners of which were often members of the above society.

Around 1800 an even more radical consolidation program was under way. By moving farms out of villages altogether, larger plots could be created, but of course the result was an even more thorough disruption of the old village way of life. The new plots were to have straight borders and as regular in shape as possible, preferably square. Rutger Maclean provided an influential example. He introduced land rents to be paid in money rather than kind, while encouraging through practical assistance the use of new forms of ploughs and other novelties. In addition, he set up schools after the Swiss educationalist Pestalozzi's system.

4.2b Direct taxation, 1720-1810: Denmark

Just as in Sweden, there were isolated attempts to introduce enclosure and reform agriculture in the first half of the 18th century. But it was only in the second half of the 18th century when the price of grain rose and the economic depression was left behind that land reform really started.[51] By 1770 half of the manorial land had been changed into the new crop rotation system. In the beginning this was a private enterprise, a technical reform carried out in civil society by means of surveyors trained in Holstein where this innovation was first introduced. Their experience was much wanted in, and since leading sections of the central administration was manpowered by large estate owners from Holstein it was a matter of routine to call this expertise to Denmark.[52] When in 1768 the government decided to sell off a great many royal estates it was considered commercially advantageous to reallocate the land in parcels enabling the modern crop rotations systems. Again, the demand for surveyors' precision measurement increased.

In 1688 only the land subject to taxation had been surveyed the result of which had been registered in books, not on maps. The geographical and economical surveying initiated in the 1760s aimed at a complete cartography. Now, the main issue was land reform and statistics, not a tax register. The precision involved may be indicated by the fact that statistics on land usage was achieved by cutting the map up into the various topographical entities (lakes, woods, for instance) and weighing small pieces of the copper-print representing each entity on a balance.[53]

The main figure in Denmark, the counterpart to Jacob Faggot in Sweden, was Thomas Bugge. From 1768 Bugge was preoccupied by the training of land surveyors. Since the cartographic project was coupled to the economic geodesy for the purpose of reallocating farm land it became crucial to instruct young surveyors about the advantages of innovation insystems of cultivation. In particular, Bugge had to convince them that the changes were advantageous not only to the landed gentry, the owners of manors, but even to the simple peasant who was suspicious of any change imposed from above. Bugge himself was a dedicated supporter of convertible husbandry and a president of the Royal Danish Society of Agriculture which had the explicit aim of modernizing agriculture.

4.3 Indirect taxation

John Brewer has shown how throughout the 18th century, indirect taxation became the main form of taxation in Britain.[54] The excise, in particular, became an organisation of extraordinary efficiency which did a lot to help finance the many wars of the period. He emphasises as a British specialty the lack of venality but also the particular form of accountancy which developed and which made it possible for the Treasury to see with clarity what the total income and expenditure of the state was. This also allows the historian to follow the British case with some ease. In this section, I will first discuss Scandinavian revenue statistics in comparison with the British case. While the Scandinavian figures are less informative than the British ones, it can at least be said that the financing of wars was central to the fiscal systems, that the state tentacles reached further and further into the nooks and crannies of society for the purpose of levying indirect taxes, that this provoked protest, and that the state promulgated standards in order to quell protests and have the operation proceed smoothly.

In Sweden, the state accountancy does not allow an answer to the question whether there was a shift from land to indirect tax. There existed names for different kinds of tax, but these did not always correspond to a fiscal category. Indirect taxation was divided up into two in linguistic use: *stora tullen* (the big customs duty) and *lilla tullen* (the small customs duty). The first was levied at sea, the second on land.

The difficulty in comparing with British figures is compounded by the fact that the customs was administered directly by the Crown in the early 18th century, whereas from 1726 onwards it was farmed out. Nonetheless, the following numbers give a very rough guide to the importance of indirect taxation in Sweden-Finland:

Overview over sample annual state revenue 1722-1806 (in silver daler coins)[55]

	Land tax	Indirect tax	Loans	Total revenue
1722	1,825,468	1,077,315		6,215,740
1734	1,958,124	1,390,300		5,604,684
1746	1,643,272	1,409,700		7,307,307
1758	2,146,042	1,660,000	8,010,000	21,374,645
1770	2,377,361	1,090,674	90,050	16,379,872
1782	780,344	549,482	3,652	3,190,655
1794	786,544	750,000	87,000	4,984,800
1806	844,882	807,651	33,333	4,097,848

For the importance of the indirect tax, the following points can be made. Excluding loans, it amounted to appr. 20% of the state's revenue, and to about 70% of the land tax. Indirect tax constituted a fair slice of the state's revenue, but not nearly as much as in Britain. Still, it was a major source of income which it would be hard to do without. Also, it proved a more flexible form of income than the land tax. From the peace time of 1722 to the war time of 1758, the land tax rose by 18%, whereas the indirect tax rose by 54%.

The geographical spread of the new indirect tax reveals the problems of collection. Town gates and to a lesser extent turnpikes provided bottlenecks for the customs officers. Town gates provided much better control than turnpikes which were more easily circumvented. In the early 1670s, characteristically, Danish provincial towns levied appr. 88,000 rigsdaler annually. By comparison, the number for Copenhagen is 69,000, and for the countryside it was only 48,500.

These numbers do indicate the geographical spread of measurement throughout Denmark-Norway in this period. And it provided growing revenue, just as was the case in Britain:[56]

Year:	1700	1706	1711	1718	1720
Rigsdaler	201,405	283,279	220,647	328,050	301,920

The Swedish Purchase and Consumption Tax Bill of 1756 reveals the level of sophistication of the tax system. Five different forms of taxation were distinguished:

1) The small customs duty was to be levied when agricultural or manufactured goods were brought to town, market or to certain designated administrative areas (including the mining districts). The tariff was paid per barrel:

wheat flour: 8 öre
wheat, rye flour, and peas: 5.5 öre
rye, barley, malt: 4 öre
oats: 2 öre

2) The port excise duty.

3) Home excise, levied on home brewing and home baking. The fee had to be paid at the local excise office. For brewing it was measured by the quantity of malt used.

4) A sales excise, to be paid by brewers, bakers, butchers, meat traders, etc.

5) Mill duty[57]

The duties were set nationally and it is to be noticed again that specifications were always in terms of a commodity, a unit of measurement and a price. The tax payer brought the commodity to a location where state-disseminated weights and measures were found. Here a measurement was performed by a servant of the state, to some extent trained and qualified by the state (the latter does not seem to be true when levying was farmed out); and the tax was to be paid in coins also disseminated by the state, the value of which was vouchsafed for by the state.

The same is true of the Danish case where the excise of the Swedes (*lilla tullen*) was copied. By 1700, the excise levied fees on almost all consumer goods, for example:

1 cartload of firewood or peat:	2 skilling
1 barrel of lamp oil:	10 skilling
1 cask of aquavit:	6 mark = 72 skilling
1 cask of beer:	4 mark = 48 skilling
1 score of eggs:	1 skilling
1 otting (16 kg) butter:	8 skilling
1 barrel of salted herring:	8 skilling
1 barrel of oat grain:	2 skilling
1 barrel of carrots:	2 skilling

John Brewer points out that the efficiency of the British revenue system relied in part on the absence of such venality, and indeed in Scandinavia, the system of farming out did not run smoothly. In the provincial town of Ålborg, a Frenchman bought the rights in 1703 to levy consumption duties. Introducing French practices he organised a search of all the town's homes, noting all stocks of foodstuffs and animals. The town's citizens were so incensed by this that they rioted. The word 'inquisition' was employed, indicating the additional problem of having Catholic Frenchmen searching the private homes of Protestant provincial Denmark.[58] The state had to intervene, but wherever possible it still sought to farm out the raising of indirect taxes, often to foreigners. The state simply did not have the means to organise such work at local level itself, and so it chose the pragmatic and imperfect solution of venality.

When the state promulgated standard weights and measures, it was explicitly in order to control abuse. The state defined ground rules for the many professions which dealt in day-to-day measurement: measurers, weighers, carriers, port controllers and firewood measurers.[59] It repeatedly decreed that standard metropolitan weights and measures be used exclusively in customs affairs. In 1687 it was decreed that all local weights and measures be destroyed and replaced with standards.[60] The repetition of such legal measures indicate that the position on the ground continued to be under less than full control.

In Sweden too a declaration was never enough to ensure compliance. For example, in 1605 the Örebro barrel was declared the standard of volume, in 1638 the Stockholm barrel.[61] But even with a standard barrel, disagreement over measurements was possible. For instance, by packing the grain one could get a barrel of higher value. The procedure of evaluating had to be regulated for higher precision of measurement and for general compliance to occur. This required controls to be disseminated throughout the network of measurement. This is the subject of the following section.

5. INSTRUMENTS, PRECISION, EFFICIENCY, CONTROL

In all cases of evaluation (money, land, merchandise), there were different levels of exactitude. In the case of coins there was weighing, hydrostatic weighing, and chemical analysis. In the case of surveying, there were rods or chains, geometrical analysis on the basis of simple determinations of angle, and high precision surveying instruments. In the case of merchandise, there were simple rods or balances used with coins, there were copies of copies of standard weights and measures, there were standards kept in the capital and finally there were high precision scientific instruments. The level of exactitude was related to the locale. After all, high precision was not always required. It was expensive, demanded great care and controlled surroundings. Low precision was cheap, quick and possible everywhere. Because of this, one can characterise the state of affairs as a geography of precision. The highest precision was to be found in the capital where standards were generated under great control. Copies of the standards were promulgated and used in the provinces under dimished control. The network of measurement and calibration became less controlled and less precise with increasing social distance from the magistrates in Stockholm and Copenhagen. This 'geography of measurement' tightened as techniques of increasing sophistication spread throughout society with the expanding administrational structure.

Looking back from the mid-18th century, Jacob Faggot, the head of the Board of Surveyors, argued that before 1680, surveying had been performed more as a craft than a science, and that it ought to be done by 'washed hands', referring to the class differences in cleanliness, and the division between intellectual and manual labour.

Faggot referred to a *de facto* professionalisation which took place between the 1680s and the mid-18th century. This section describes the professionalisation between 1660 and 1800, both in terms of the creation of a monopoly, in terms of training, and in terms of control. The historical sources allows this to be illustrated particularly well for Sweden-Finland, which thus takes pride of place in this section. The process of professionalisation was accompanied also by the growth of a market in scientific instruments which will be described subsequently. Much of the instrumentation explicitly addressed the problem of striking a balance between precision measurement and facility of use under varying conditions, and this I will illustrate with examples. Thus both professionalisation and the development of scientific instrumentation concerned itself with the geography of precision measurement. After documenting the professionalisation of measurement and the use of scientific instruments, I will argue that coinage is an intrinsic part of the development of a state network promoting stable value through precision measurement and promulgated standards described so far.

5.1 Professionalisation of measurers

In mid-16th-century Sweden, bailiffs, the county chief and a committee of local peasants carried out land evaluation. Local men or boys were employed measure with rods. Complaints from peasants about unjust taxation persisted throughout the century. A Board of Surveyors was set up in 1628 specifically in response to the complaints, and with the aim to make taxation more just and thereby facilitating its collection.[62]

The new surveying was to be coordinated by the prefects. All measurement protocols were to be submitted to the *kammarkollegium* Area and quality of the soil (measured by yield of number of haystacks) was to be reported along with the owner's name and number of grown-ups living on the land. A land register was to be collated from this. Initially, maps were not drawn.

In the last decades of the 17th century, when the *reduktion* (the state confiscation of nobility owned land and its subsequent parcelling out to soldiers) was at its peak, surveys were carried out all over Sweden. New regulations were introduced: in 1687, the king prescribed that surveyors take an oath as qualification for this kind of work. It was also decreed that houses and properties be searched as part of the evaluation, and that an inventory be established. Surveys, and by extension evaluations, were controversial, so that four additional local civil servants were to participate. A report of measurements, evaluation, and drawings was to be submitted to the *kammerkollegium* for each unit of land surveyed.

We have repeatedly seen that several trustworthy individuals apart from the surveyor had to take part in the measuring process for the result to be trusted. Who measured mattered and the regulations were getting increasingly explicit about whom to involve. In addition, the question of measurement technique also gained in importance. Some of the early instruments were very simple. Mostly, ropes were employed. They had the advantage over rods that they were easier to transport. In the mid-17th century, at the very latest, the measuring board associated with Johann Prætorius of Nuremberg was introduced into Sweden. A textbook from 1670 argued that a rod is more reliable than a rope, because the latter shrinks when wet, expands in the heat, and markings on it tend to disappear, whereas the rod does not change. It also pointed to the importance of markers in the field and taught to calculate the area of an irregular field by dividing it up into a number of squares. It is, however, only towards the end of the 17th century (when the *reduktion* and the parcelling out of land for soldiers were under way) that geometrical instruments were used. (Geometrical was the term employed to distinguish the instruments measuring angles from rods and chains measuring lenghts.) These instruments involved more complicated geometrical calculations.[63]

In 1683, Axel Oxenstierna (a nobleman, and Faggot approved of him) was appointed head of the Board of Surveyors. How to measure now became an explicit issue. A number of initiatives were taken in the next dozen years. Surveyors were to be examined and licensed. All rods, chains, and ropes were to be calibrated against the Stockholm unit of length. Also, the intellectually more demanding instruments to be employed by 'washed hands' such as astrolabes, quadrants, compasses, spirit levels, proportional circles (invented by Galileo), rulers and similar instrument were to be introduced. A special tool, the 'linear transporter' was to be disseminated, which enabled the conversion of a map with one scale into one with a different scale. The commensurability and convertibility of scale was important in map making just as the commensurability and convertibility of units was in, say, weighing.

Just as tax collection was farmed out, so was surveying. Surveyors were to purchase a standard with their own money which in fact they were reluctant to do. The head of the Board of Surveyors had to explain the request for further money from the *kammarkollegium*: Surveyors still employ rods and ropes despite regulations to the contrary. 'But this technique is much inferior to those of the mathematical sciences. The latter can not fail, whereas the former causes trouble especially with height differences and large distances.'[64]

As a result of Scanian peasant resistance in the 1750s and 1760s, a new instruction for surveyors, replacing that of 1725, was introduced in 1766. It prescribed in much greater detail just what was to be measured and reported back to the central administration. In addition, the peasants involved had the right to appoint independent 'reliable' (what this implies is not specified) observers when the quality of the soil was to be evaluated (except when the evaluation was simultaneously done for the purposes of determining the tax level). These independent observers were to try to mediate and resolve any dispute, before the case was referred to local courts.

Consolidation now took off and surveyors were stretched to the limit. According to one prefect in 1769, the new plots were mostly embanked, cleared of stones and crop rotated, and the yield increased several times over. Despite the increase in consolidation, it was not universal. Where agriculture provided a side income, consolidation seemed too expensive. And generally, misgivings about the redistributions remained.

By 1783, a new surveyors' instruction specified further the process of evaluation of land quality. Fields, meadows, newly cultivated land, forests etc. were now to be evaluated separately. Within each kind there were several gradations for the possible value. At the same time the ways in which compensation for exchange of superior against inferior land could be paid (manure, labour, or money) was determined.

What applies to surveying applies equally to weighing. In fact, surveyors were put in charge of weights and measures. The local diversity in measuring practice which surveyors faced was great. As Witold Kula has described in great detail for Poland and France[65], not only were weights and measures different in different localities, but also the practice of topping up after weighing or measuring was common. The practice of exchange was not just a question of determining the equivalent value of two wares, or that of a ware in money, but an almost ritualised encounter. In Sweden, topping up could sometimes add 20% or more to the measured volume. Kula has shown that within local communities everyone knew the rules of the game, but with increasing circulation of traders and wares, the local varieties became increasingly problematic. The state, which was attempting to coordinate the activities across many local communities had an incentive to standardise. For instance, in 1638 it was prescribed that in Sweden the unit barrel was to refer to contents of a physical barrel without top and without shaking or pressing the contents.[66]

In the 1660s Georg Stiernhielm was appointed to reform the weights and measures. Stiernhielm had learned connections: in 1669 he was elected Fellow of the London Royal Society, and he advocated the introduction of the decimal system into Sweden. Prompted by Stiernhielm, weights were now for the first time separated from the issue of coinage (see also section 5.3). An inspector was appointed by the *kammarkollegium* to oversee all large balances and weights of the realm, except for the small ones intended for gold and silver which continued to be under the auspices of the minting examiner (*riks-guardien*). In 1665, a declaration was posted that only certain balances and weights were to be employed, but no additional hands were employed to test existing instruments. Calibration was to be the responsibility of the general administration, which was naturally busy with other matters.

Danish weights and measures were reformed in 1683, when a regulation was devised in consultation with Ole Rømer. Rømer had intended all weights and measures to be reducible from one standard length. Standard weights, for instance were given by cubic vessels of standard length, containing pure water. But the reality of weights and measures remained a far cry from Rømer's grand plans for order.[67]

The 1683 regulation prescribed national standards and was followed up by supporting regulations intended to ease through the introduction of the new standards on a national scale. In order to ensure that weights and measures were indeed standardised, the Copenhagen magistrate was given monopoly of their sale. But craftsmen resisted this monopoly vigorously also after Rømer succeeded in wresting the monopoly from the magistrate in the late 1680s. When it was clear in 1687, that old measures were still being used, their destruction was prescribed but to no avail. Eventually, the king had to interfere in a dispute which dragged on until 1698 between the pewterers, the magistrate and Rømer. A new regulation now gave the Copenhagen magistrate and 4 provincial towns monopoly on sale of standards.[68]

The difference between legal prescription and actual practice remained, but it was reduced.[69] From 1698 onwards the Copenhagen magistrate began to take advantage of its monopoly. Accounts reveal that the magistrate made a nice profit over the cost of labour for cooper and smith when selling weights and measures. Citizens could also bring their weights and measures to the Town Hall for calibration against a fee.[70]

In the 18th century, the geographical spread of precision measurement continued at a quickened pace. In the early part of the century peasant grievances about the abuse of weights and measures increased in Sweden-Finland. In response, the parliament debated the pros and cons of a corps of well-qualified calibrators (*justerare*) spread out over the entire realm. In 1735 the government decided to add weights and measures to the responsibilities of the Board of Surveyors. Market towns were to have a calibrator taken under oath, who also had to pass an exam at the Board of Surveyors. Calibrators were to be remunerated through fees, and it was to become compulsory for traders to have their instruments calibrated.

Measurement of barrels was a crucial task. Previously, foreign measuring rods had been used to evaluate the contents of barrels for importation, but they were now prohibited and to be replaced with Swedish ones, such as the ones constructed by Daniel Ekström. The private company to which the customs business was now farmed out (*Generaltullarrendesocieteten*) paid for Faggot to produce a thesis on *Volume measurement, or observations on the art of measurement for the deduction of the contents of various vessels in Swedish units*. An abbreviated version was printed in the Academy's transactions (cf. section 1).

In the 1740s, the task of the calibrators was to some extent coordinated with local prefects. In the 1750s, just in time for the expensive war with Prussia, the enforcement of the laws was aided by the specification of punishments. All instruments (barrels, mugs, rods, balances, etc.) were to be calibrated and hallmarked. Any trader possessing instruments not yet hallmarked was to have them confiscated and fined. Anyone using the wrong measures was to be put in the stocks for an hour, and to receive ninety beatings. Calibrators faking a hallmark were sentenced to death. Anyone changing a hallmarked instrument was fined and 'dishonoured'.

In the same period the problem of wear and tear was regulated. Weights in particular were known to change with use, and the customs asked Faggot for advice. As a result the Board of Surveyors came up with guidelines for corrections to wear and tear. Also, topping up was prohibited while exact recompense was prescribed in some cases.

Because of these practical organisational problems, the instruction of 1766 reiterated the need for the introduction of calibrated instruments. Now, the practice of surveying was prescribed in even greater detail. There were rules for the setting up of corner stones, and rules for dealing with the topography of the land. Reconnoitering in advance of the actual measurements was emphasised. And there were further rules for the way in which maps were to be drawn and information to be recorded. In the winter and in all hours of leisure, the surveyor was to make fair copies of all his drafts. The kind of paper, pen and colur to be employed was prescribed too. In the era of land consolidation, the instructions were shorter lived, and a new instruction was issued in 1783. One could summarise it as providing even greater discipline without any technical novelties.[71]

In the instruction of 1783, the right of anyone to become a calibrator provided he (gender used advisedly) passed an exam, came to an end. Now, calibrators had to be appointed by the local administration. In the 1790s, the tasks were routinised. For instance, in Stockholm, every month belonged to the calibration of instruments employed in particular trades. In January all butchers, fish mongers and iron mongers were to have their instruments calibrated. In February it was the turn of brewers, bakers, distillers, and vinegar makers, and so on. There were also attempts to spread the control of weights and measures into some walks of life which failed. For instance, bottles could not be calibrated, since the glass blowers were not capable of blowing sufficiently regular bottles.

Still, there can be no doubt that the Swedish state became ever more capable of collecting indirect taxes. The numbers given in the table above which reveal a decline in the total income from indirect taxes from 1758 onwards will be due to the political undesirability of stifling trade and the ceasure of an acute public service borrowing requirement from an expensive war. They do not reveal the decline of the importance of standards or of the efficiency of the weights and measures department of the Board of Surveyors.

In fact, the question of efficiency became explicit in the course of the 18th century. How much work could a surveyor be expected to do in a given time period. In other words, some means of keeping a check on surveyors' diligence was desired. It was generally agreed that something in the order of 80 farms a year per surveyor was acceptable, with the provision of course that some terrain was too difficult to make this a reasonable target for one surveyor.[72]

5.2 Instruments

It seems that until the 18th century, most scientific instruments had to be imported into Scandinavia from abroad, usually from Germany. In Denmark, the standards of 1734 were made by an instrument maker by the name of Carl von Mandern, probably a German. But during the 18th century, a few instrument makers made it to some level of prominence, first of all in Stockholm. It seems that scientific instrument making was much more highly developed in Sweden than in Denmark, although it might be that this impression is enhanced by the much greater amount of work done by Swedish historians of science than by Danish ones.

Swedish instrument making reached a high international level in Faggot 's protegé, Daniel Ekström. In 1750 Faggot complained that 7 other Stockholm instrument makers supplied inferior ware. This indicates that there was a market for cheaper, less accurate instruments too. One of these instrument makers, who had been Ekström's apprentice, produced dividers, rulers, and compasses; again surveying instruments.[73]

Whereas only two instrument makers based in Copenhagen have been identified (Ahl and Bidstrup, see below), many more Swedish instrument makers are known[74]:

Akrel, Fredrik (1748-1778); Engraver to the Swedish Academy of Sciences 1778-1804. Terrestrial and celestial globes

Apelquist, Carl (c.1749-1824); Gunfounder at Norrköping, Sweden. Studied with Jesse Ramsden and William Fraser in London for several years. Returned to Sweden in 1778. Founded an engineering workshop in 1785 for the manufacture of tools and machines and for cutting wooden screws and stamping metal buttons. Set up a foundry just outside Stockholm (Marieberg).

Collin, Gabriel (1761-1825); Employed at the Council of Mines, Member of the Swedish Academy of Sciences from 1812. Studied mining engineering at Uppsala University from 1782. Founded an optical shop in 1790. Took over the shop of the Swedish Academy of Sciences from Henrik Holmbom.

Ekström, Daniel (1711-1755); Instrument maker to the Board of Surveying (1735-); Member of the Swedish Academy of Sciences (1742-). Apprenticed in

1727 to Petter Rosenberg, apparently the only instrument maker of note in Stockholm. Went to Uppsala in 1733 to attend the Laboratorium mathematico-oeconomicum. Travelled to London and Paris 1739-1741 on a Government grant. Followed Desagulier's lectures in experimental physics, visited Greenwich Observatory, spent much time in the workshop of George Graham. Set up instrument making shop in Stockholm in 1741. Achieved reputation for high quality instruments even outside Sweden. Granted an annual salary of 6000 Rigsdaler from 1751 on from the funds of the Board of Commerce and given the title of Director of Instrument Making, on the condition that he took on apprentices which were examined by the Swedish Academy of Sciences. The best known of these are: Johan Ahl, Carl Westberg, and Johan Rosenberg.

Ernst, Petter (1714-1784); Clockmaker. Started a business in Växjö c. 1734. Moved to Stockholm in 1754 on the suggestion of the Swedish Academy of Sciences. Put in charge of the clock factory Stockholms Manufabrique. The shop produced altogether about 1150 clocks and watches, or about 25 a year.

Hasselström, Johan Gustaf (1742-1812); Maker to the Swedish Academy of Sciences 1777-1812. Apprenticed to instrument maker G. Duhre in Stockholm. Opened his own business in 1775.

Hoffvenius, Johannes (-1685); Instrument maker ot the University of Uppsala (1663-); Granted exclusive rights to making scientific instruments in Uppsala and Stockholm; later became surveyor and finally Surveyor General

Holmbom, Henrik (1752-1793); Optician. Studied for some time at the University of Uppsala, and after 1775 worked at the Swedish Academy of Sciences' optical workship taking charge of it from 1784. Had two apprentices.

Hultberg, Olof (c.1708-1743). Studied at Uppsala from 1723. Made an air pump and hydrostatic balances for Mårten Triewald's lectures in experimental physics. Worked in the 1730s as instrument maker in Uppsala.

Lehnberg, Carl (1725-1768); Controller at Kungsholmen's Glassworks, Stockholm (1747-); ran the optical part of Ekström's shop in 1755; member of Swedish Academy of Sciences (1756-). Studied optics at Uppsala University for three years around 1745 under Samuel Klingenstierna; two-year study trip abroad (unspecified). Made first Swedish achromatic telescope in 1760.

Rosenberg, Johan Petter (-1777); Maker to the Swedish Academy of Sciences (1764-1777). Apprenticed to D. Ekström. Made many instruments for the physicist J. C. Wilcke

Thelott, Olof (c. 1695-1713); Clock maker and engraver. Instrument maker to Uppsala University in 1711; demonstrator at C. Polhem's technical school in Stockholm in 1713. Son of Philip Thelott, Swiss born engraver and clock maker, summoned to Sweden by Professor Olaus Rudbeck in the 1670s.

Westberg, Carl (1720-1769). Apprenticed to D. Ekström. Opened on his own in Stockholm in the 1750s. From 1756 subsidised by a government grant. Made microscopes, surveyors' instruments and sundials. Left the shop to his former apprentice and foreman G. Duhre in 1763.

We know most about Ekström, by far the most successful Scandinavian instrument maker in the 18th century. In 1735, he was made instrument maker to the Board of Surveyors and he opened a small workshop in Stockholm. In 1739, Faggot managed to elicit a large stipend from the *kammarkollegium* for Ekström to go on a European study tour. He spent a year and a half in Paris and London, some of it with the renowned George Graham.

After returning in 1741, Ekström set about supplying the Swedish market with scientific instruments. Some insight into the nature of the market can be had from a memorial by Faggot which lists the instruments which Ekström produced:

1) mobile quadrants with tubes and micrometer
2) transit instruments
3) large tubes with and without micrometer
4) reflection quadrants
5) Gunther's scales etc.
6) levelling instruments of all kinds
7) astrolabes with various accessories
8) dioptric rulers with and without tubes
9) proportional circles with all kinds of gradations
10) transporters and other gradaded instruments of variouys composition
11) compasses with various accessories
12) geometrical and trigonometrical scales
13) measuring rods for all sorts of use
14) quadrants for the aiming with guns and mortars
15) drawing aids? (ritcirkelbesticker på åtskilliga sätt)
16) copying instrument for the up- adn down-scaling of maps and drawings
17) all kinds of beam compasses with pen and scales
18) sundial (?)
19) fine analytical balances with appropriate weights
20) hodometer and measuring chains
21) reflection telescopes
22) image-reflector and opera glasses
23) small and large images
24) microscopes of various constructions
25) camera obscura and other optical machines
26) various airpumps
27) compressors

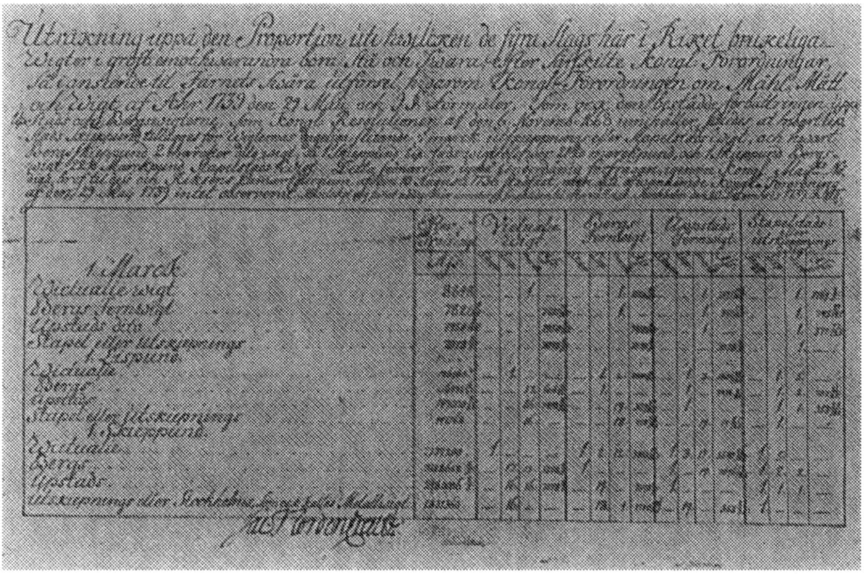

Illustration 3.

28) hydrostatic balances
29) barometers

A great many of the instruments are for the purposes of surveying: 1, 2, 5, 6, 7, 8, 10, 11, 12, 13, 16, and 20, and possibly some of the other categories too. Ekström supplied instruments to members of the Royal Swedish Academy of Sciences, of which he himself became a member in 1742, three years after its foundation. But one can safely conclude that the reason there was a skilled instrument maker in Sweden was to the state's demands for precision measurement (employed as he was by the Board of Surveyors), although Faggot does not tell us anything about the numbers of instruments produced. We do know that Ekström was able to supply the instruments at a price below the foreign competition.[75]

It was also Ekström who in 1739 produced a standard balance with weights, against which all others be calibrated. It came with a table to be used for conversions (Illustration 3: Uträkning uppå...'). The table refers to the troy unit of weight (used by the Dutch and German traders), the victuals unit, the *Bergs Jernwigt* (mining area unit), *Uppstads Jernwigt* (market town unit), and finally the *Utskieppnings Wigt* (embarkation unit). The necessity of this table was due not only to

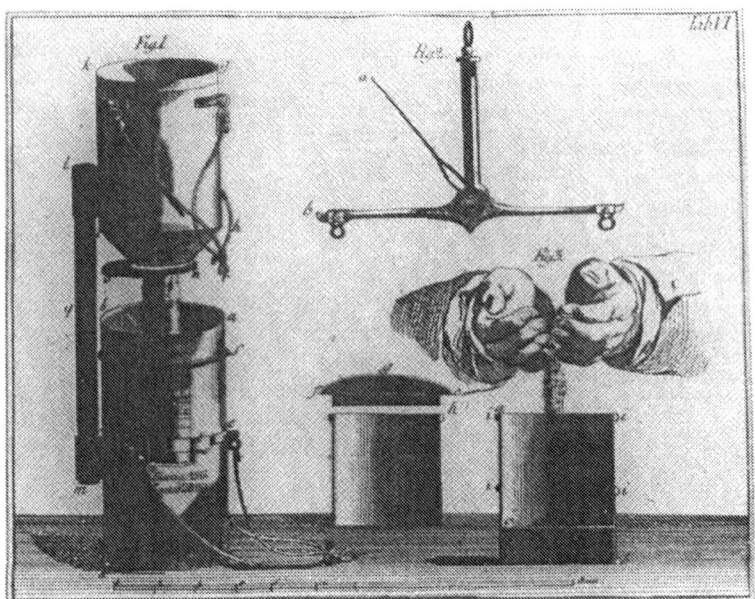

Illustration 4.

different units employed for different wares, but the use of different standards of metal weight in different locations. The purpose was to support the transport trade from the mining areas to the ports on the coast, by rendering units progressively smaller. Therefore, a higher value on the same amount of metal was to be had along the route. The names of the units were the same everywhere (skippund, lispund, marck, and ass), but not their size. As Kula has shown, this was common all over Europe.[76]

Another instrument devised by Ekström, the grain tester, is very informative about approaches to measurement (Illustration 4: 'Tab VI'). Its purpose was to regularize the measurement of the quality of grain seed by comparing its weight and volume. It was not designed only for official use; Ekström emphasizes that any tinsmith could make it, and he recommended its widespread adoption. It would improve accuracy of testing even when reproduced with only moderate tolerances, because it required no close measurement of volume; in conjunction with an accurate balance and weight, it allowed reliable comparisons between grain samples.

The way in which the grain is poured is critical since it affects packing; in mill practice it was considered cheating to pour grain into a barrel from above the hip. Similarly, when testing samples of grain, seed would be dropped carefully into the container from cupped hands as shown. Again, the way of pouring the seed could be employed to pack the seed differently. Ekström's instrument is intended to eliminate the uncertainty of the use of hands in the normal measuring process (Fig. 1 on the left hand side of the illustration is Ekström's instrument, the hands on the right simply indicate the usual procedure which he aims to eliminate). The container *ab* has an adjustable floor *c*. The container *ab* is now put underneath another container *ghik*, which has a hopper at the bottom that can be opened and closed. With the hole closed, the seed is poured into *ghik*. The rod *lm* determines the height between *ghik* and *ab* and can be set to a standard distance. Thus, to whatever capacity the container is set, the distance between hopper and floor remains the same. By opening the hopper, the seed then falls into *ab* in a standard way unlike the pouring from cupped hands.

In preparation for the measurement, the seed was to be dried to prevent the water content from influencing the measurement. Ekström suggests standard ways and lengths of time for the drying process. Once the grain samples to be compared are equally dry, a known weight of the first grain type is poured into the container until the grain fills the container to overflowing, and the surplus is levelled off, flush with the rim, using a specially constructed tool which Ekström describes in some detail. The container is then emptied, without changing its capacity; grain of the second type is then also poured in, until it also fills the container exactly to the rim. It is interesting that Ekström is not satisfied to use a container of fixed capacity. Probably, this is so that the container can be adjusted to the standard weights for each kind of grain.

With many grain testers of this kind, the quality of seed could be ascertained everywhere. Ekström referred to the need for distributing grain testers by mentioning the most recent Royal decree on barrel measures. The state actively supported this kind of work; in fact Ekström performed the experiments on the order of the *kammarkollegium* at the royal grain depot. Ekström also claimed that the instrument would help to mediate in international grain trade, referring to Dutch standards and the requirements of the Dutch market.[77]

In Denmark-Norway there were no prominent instrument makers until the end of the 18th century. Before that time, high precision instruments had to be imported, indeed some of them were Ekström's. But less well known instrument makers supplied standard weights and measures to the Copenhagen magistrate. These standards were used to calibrate all weights and measures brought to the Town Hall, and a fair amount of new weights and measures were sold too.

Accounts kept in the Copenhagen Town Hall enable an estimate of the calibrations carried through and secondary standards sold, thus giving an indication of the geographical spread of calibrated measurement in Denmark-Norway of the time. The magistrate's Chamber of Calibration received just over 1,637 rigsdaler in 6 months of 1772 for sold weights and measures and calibration fees. Calculating from the prices and fees given this would correspond to, for example, the sale of 96 fitted salt barrels, 96 tin pots, 96 sets of 10 *lispund* weights, 96 sets of 1 *lispund* weights, plus the calibration of 96 salt barrels, 96 oil barrels, 96 tin pots, 96 tin jugs, 96 iron rods (1 *alen* long), and a 10 *lispund* metal weight. And this activity increased: in 1794, the income for calibration and sale of calibrated weights and measures had increased to just over 2,795 rigsdaler.[78]

A publication in the Transactions of the Copenhagen Academy illustrates the production of a metropolitan standard aiming to improve the levying of an alcohol excise.[79] It addresses the questions of precision and reliability in different locales. It was entitled 'On the means of examining, testing and evaluating all spirits traded, both in terms of their measure and their quality, etc.'.[80] The author, Franz Heinrich Müller, stated that the Customs Office had set him the task of making an instrument for the purpose of assessing alcohol content[81], in order to 'measure their value in money'[82], while also emphasising that every trader can use it, since he must know the value of merchandise to avoid harming either himself or his customer and retaining the faith of the public.[83] The parallel with Ekström's grain tester is extensive: precision measurement employed for the purpose of evaluation; the making of a standard instrument initiated and paid for by the state but intended also for the market.

Müller refers to the measuring rods of standard size usually employed to measure the content of barrel by holding them diagonally within the barrel (as mentioned above Faggot had published a paper in the Swedish Academy's Transactions working out the geometry required to establish volumes from the exterior dimensions of vessels - this is the kind of problem Müller had in mind).[84] The much more taxing problem was the alcohol content.

His solution was a silver rod which was dipped into the liquid in question (see Illustration 5: Scala for en Brændevins Pröver). The portion submerged indicated the alcohol content, and the silver rod was to be given markings denoting alcohol content. These markings were applied in a metropolitan laboratory where the purest water and alcohol available were mixed in varying proportions and the dip of the silver rod tested. The rod is depicted on the left of the illustration and the columns on the right refer to I: the total quantity of liquid; II: the mix of water and spirit. The original calibration of the rod was performed at 0 degrees Reaumur (already fairly advanced, requiring control of temperature and a thermometer calibrated on the Reaumur scale). The calibrations on the rod with the numbers in column III, referring to the quotient of water and alcohol. However, this quotient was different for moderate and hot weather (given in columns IV and V, respectively). These quotients were also tested at the original calibration.[85]

Müller emphasised that other methods of alcohol content were less useful: distillation was too expensive; burning required absence of wind; a hydrostatic weight

Illustration 5.

was fragile, slow and required to be used indoors; judging by the bubbles created by shaking was unreliable.[86] His rod thus travelled better than any of the rival candidates. Still, he did emphasise that the rod was subtle and must be protected from knocks and falls with a padded container.[87]

The rod could be used to detect fraud almost instantly. One fraudulent technique was to heat the liquid just before the measurement was to take place, and temperature must thus be carefully checked.[88] The trust inspired in the value of the merchandise by the spirit tester was to provide savings in transport, insurance, and other costs.[89]

This silver rod spirit tester thus illustrates very clearly the issue of calibrating an instrument which can be used with ease out in the field where measurement and evaluation takes place. It also shows the problems of control, authority and fraud.

Towards the end of the 18th century, many such instruments were in demand. The demand for a local precision instrument maker thus became acute. The surveys required good sextants, mural quadrants, geographical circles, levelling instruments, compasses, chains of measurement, good quality paper, etc.[90] Customs and excise required a variety of instruments such as Müller's silver rod. But no instrument maker capable of producing the kind of quality which was the norm in European centres lived in Copenhagen, nor were the requisite skills to be found with the already present crafts. It required that someone gathered the experience abroad and returned to set up shop in Copenhagen.

There were only two instrument makers of renown in Copenhagen in this period: Johan Ahl and Jesper Bidstrup. Ahl had been apprenticed to Daniel Ekström in Stockholm and after running up a debt, he escaped to Copenhagen where he was suborned and set up his own workshop in 1795.[91]

When Thomas Bugge was put in charge of mapping Denmark-Norway, he selected one of his students, Jesper Bidstrup, and sent him to London as a trainee in the instrument making profession.[92] Bidstrup spent seven years in London. Bugge took advantage of his international network by recommending his young protegé to Joseph Banks. This undoubtedly assisted Bidstrup in breaking into the closed world of London scientific instrument making. London instrument makers charged a high fee for apprentices and their products were protected by the Tool Acts. Access to Jesse Ramsden's famous dividing engine, a machine tool employed in the making of many scientific instruments, was only open to British members of the trade. But with his connections, including masonic brethren, Bidstrup eventually was apprenticed to a supplier to instrument makers. As a foil, Bidstrup set up shop at Leicester Square and published a sales catalogue. He shipped a sextant to Copenhagen and managed to copy essential parts of Ramsden's dividing engine, including a screw cutting lathe. He also bought parts of a glass grinding machine and a tube machine. In order to circumvent the Tool Acts, he shipped the various bits in different consignments to different destinations and eventually Bidstrup himself returned to Copenhagen. As a result some degree of scientific instrument craftsmanship was imported into Denmark.

5.3 Coins as measure of value

The history of coinage is not usually seen as a part of the history of science. In fact our disciplinary boundaries normally carve it up into economic history, numismatics, and history of technology. This fact makes coinage a very tricky subject to tackle historically. The literature is diverse and largely incompatible. I wish to argue that the history of coinage is central to the history of science in the service of the state. (Notice for instance, that it was only with Georg Stiernhielm in the 1660s that the issues of coinage was institutionally separated from weights and measures in Sweden, as mentioned in the section on professionalisation above.) Coinage was always crucial to the collection of taxes, as far back as Roman times.[93] The value of coins was forever in dispute, and there were repeated attempts to standardise them. Furthermore, the precision of the value of coins continued to be an issue and the state mostly took great pains to keep coins as uniform and standardised as possible. The process of minting was highly specialised, involving both precision weighing, chemical expertise, and extreme control. The value of the coin produced under these circumstances was vouchsafed for by the image of the sovereign on the coins themselves, very much like the standard weights, measures, and rods. There was also a geography of measurement with a hierarchy of precision, the highest precision achieved at the mints, the lowest in a provincial marketplace. Measurement and evaluation is thus intimately involved in coinage and many of the problems are the same as in weights and measures.

The main techniques for testing gold and silver content in coins were widely disseminated all over Europe even in the Middle Ages.[94] Weighing was one way of assessing a coin's value, but coins could be rendered heavy by an admixture of lead. A more sophisticated measure consisted of the hydrostatic weighing, that is weighing consecutively in air and water which, as we would now say, detected lead by determining the specific weight of the amalgam.[95] The sophisticated test which was known to goldsmiths all over Europe in the Middle Ages and remained an integral part of minting practice throughout the period concerned here was a full-blown chemical analysis. Towards the end of the 18th century, the trade secrets of coin assaying were published in book form.[96] The sample was placed in a small container made of clay. Aqua regia was added and the container was heated to a high temperature in an oven. The impurities were absorbed into the walls of the container leaving the noble metals to be weighed.

These analytical and measuring practices spread throughout society wherever money changed hands. Simple, cheap balances were more widely employed, and many traders would employ hydrostatic balances (especially towards the end of the 18th century), whereas the costly and time consuming full-blown chemical analysis was restricted to only a few places, such as mints.

The value of coins was vouchsafed for by the organisation of the mint in a similar fashion to the organisation of a laboratory: restriction of access, use of sophisticated scientific instruments and accounting practices, enabling the sovereign to keep some degree of control over work in the mints without actually having to be present at the time of minting. Furthermore, the work of the Master of the Mint was surveyed by a Guardian, employed for that purpose. The regulations and control techniques at the Mint got ever more sophisticated in the 17th and 18th centuries. At the Royal Danish Kongsberg Mint, the instructions for the Master of the Mint and the Guardian in 1628 and 1629 were very brief by the standards of a century later (the text is roughly a fourth in length). The Guardian of 1629 had mainly to observant and diligently assay the amalgam before it is struck.[97] The instructions for the Guardian in 1730 consisted of 14 paragraphs. He now was obliged to take samples several times during the minting process and it was prescribed who was to be present. He was then to analyse the metal contents in his own laboratory. The locking up and sealing of samples, the accountancy and report required were much more detailed. He was now also allowed to calibrate samples brought to him against a fee, while remaining under his oath of fidelity to the king, thus very much resembling the role of the Stockholm and Copenhagen magistrates.[98] Again, trust, authority, and detailed regulation combine to vouchsafe for a stable value.

The organisation of the Swedish Mint was no different. A manuscript which has been dated to 1720, refers to the practice of minting, the control functions of the Guardian, and the bookkeeping procedures in virtually the same terms as the Danish one of 1730.[99] In it, the authority of the sovereign is addressed:

> Those who counterfeit should receive harsh punishment, not just for reaping a large profit but especially for profaning so shamefully the royal prerogative... Noone should disrespect the royal mint; rather one ought to respect the country's honour and the royal prerogative, not tread it under foot. The ancient Romans considered throwing a coin with the sovereign's image into the latrine a *crimen læsæ Majestatis*.[100]

The state attempted to counter coin clipping and the filing off of metal dust with harsh punishments but also by setting limits for the acceptability of coins. They were to possess weights within particular limits. If found too light, they could not be accepted as tax payment. Similarly, the content of precious metal in the coins had to be within certain limits. This was called the 'remedium'.[101] The precision with which the mint produced coins within the limits was extremely important. If the content of precious metal was too high, the coins would disappear from circulation and be melted down; if it was too low, it would not inspire trust (especially as the state itself might not accept it is tax payment) and it would fall below its stipulated value causing inflation. If coins were produced with too large a spread in the precious metal content, the more valuable ones would disappear. This was called 'seigern' in German and the process was denoted 'Gresham's law'.[102] The Crown thus had a great interest in the production of coinage with very precise contents and weights and in preventing tampering with coins, both for the purposes of having money with which tax could be paid, and for the purposes of enabling trade and avoiding inflation.[103]

The social organisation of the mints and their use of scientific instrumentation was a result of this concern of the state. To glimpse the kind of precision work performed in a mint, we can inspect an 1806 inventory of the Kongsberg mint. From it we can see that precision measurement took place here, along with chemical analyses of various kinds. One room was called 'laboratorium', where a 'Probeer-Vægt' (assaying balance) and two 'Proberovne' (chemical ovens) were found. Next door was an iron balance with copper bowls and weights of various sizes. In another building was a room with two rough, five precision, and two very high precision balances. Many rooms contained ovens and various kinds of tools for handling materials in the ovens. There was also a 'Kontoir' (comptoir or office) where containers with locks were kept. Many doors had locks bearing witness that this was a place with highly restricted access just like modern laboratories.[104]

The publications of the Academies also dealt with precision work in minting, as mentioned above. For example, Müller (the same person who devised a spirit testing instrument) depicted many tricks of the trade, analysing various aspects of minting, such as the dimensions of the oven, the best means of producing vials and containers, the impact of various ways of stacking the coal.[105] The analysis was explicitly intended to improve precision.

DISCUSSION: MORALITY, MONEY, MODERNITY

The aim of precision measurement as exemplified by the case of minting was to generate stable values upon which everyone could agree. The same goes for weights and measures where buyer, seller, tax collector and tax payer agree upon the measure (and thus the value) to be handed over. This would facilitate tax collection and exchange of goods. As Witold Kula has pointed out, morality of exchange is an integral aspect of weights and measures:

> In the Books of Moses, which constitute a code of social conduct hedged with sacred sanctions, the norms relating to measures are still put literally. Thus we read: "Ye shall do no unrighteousness in judgment, in meteyard, in weight, or in measure. Just balances, just weights, a just ephah, and a just hin, shall ye have." Or again: "Thou shalt not have in thy bag divers weights, a great and a small. Thou shalt not have in thine house divers measures, a great and a small. But thou shalt have a perfect and just weight, a perfect and just measure shalt thou have: that thy days may be lengthened in the land which the Lord thy God giveth thee." The religious sanction against metrological transgressions is therefore death.[106]

Many works on the history of standardisation emphasise this role of standards as protection against cheating. As we have seen repeatedly, in Scandinavia also, the sanction against metrological transgression was death.

Kula has also shown that we can not simply view this as the state introducing fairness. New weights and measures were resisted all over Poland whenever these were introduced to replace local kinds. Kula shows that there was a great variety of measuring vessels, even for each kind of commodity. For instance, barley and rye would not usually have the same measure, and addition, some types of grain were measured with top (the top created when pouring the grain into the vessel) while for others the top would be levelled off carefully. What seems bewildering was not so to the locals. Each way of measuring was well known to all, including the amount of top which should be added for each type of grain. In fact, locals often resisted the introduction of new weights and measures, precisely because their lack of familiarity with them opened up possibilities for fraud.[107] This is also the case for reactions to the work of surveyors, who after all determined the value of land by particular non-local standards and precision measurement. Polish 'peasants would often invoke the devil to lay his talons on the land surveyors.'[108]

This resistance to imposed weights and measures seems to be very common. The introduction of new metrological regimes upset the routines and so actually enabled fraud, at least in the short term. Peter Linebaugh has graphically described how metrology helped to eliminate the perks of workers, such as the London harbour proletariat - perks which had been a normal part of the remuneration. Metrology thus effectively lowered the pay of the lowest paid.[109]

Weights and measures impinge directly on the routines of exchange and as such they are moral; they constitute a part of the ethics of fair trading. What happened in the period from 1660 to 1800 was the creation of a state infrastructure which was sufficiently large and powerful to reach down into local places of exchange and to have a say in the fairness of evaluation there. The growth of the state, its science and metrological regimes had a significant impact locally.

Precise evaluation was a part of a development which changed the social fabric markedly. In this there were winners and losers, and hence the tensions were significant. The elimination of local weights and measures undoubtedly reflects the breaking up of local communities as trade increased and as state control at local level increased. Local self-sufficiency was on the wane in the period concerned, while payment in kind was replaced by payment in money. The administration of the fiscal-military state grew and the agricultural revolution integrated rural practices with urban tool production and capital.

This does sound like some form of modernisation, although it is always difficult to argue for a general process which all cultures go through.[110] Nonetheless, a few comments are in order. Karl Marx, Max Weber, and Georg Simmel have all argued for some form of relation between the role of money and rational thought in the modernisation process. Marx's condemnation of money and market exchange does contain some nostalgia for the local community where production was for use and did not become the basis of systematic unfair exchange. The new modes of production denied those moral 'bonds which unite men one with another', as Émile Durkheim had it. Weber's concept of a development towards a rational society involved monetization and bureaucratisation. For Simmel, it was under the aegis of money that the 'modern spirit of calculation and abstraction has prevailed over an older world that accorded primacy to feelings and imagination'.

In contrast to Marx, Simmel argued that money had also an enabling effect: while a money economy does erode older solidarities it simultaneously promotes a wider and more diffuse sort of social integration. In the case of barter, trust is confined to the individuals engaged in the transaction; whereas monetary exchange requires and bolsters a trust to an enormously expanded social universe.

It does seem that once a metrological regime is established, exchange is facilitated because each partner in exchange can trust the state authority to force the other to use only the prescribed form of evaluation. We do not now need to check whether the petrol station actually sells us petrol in the litres signalled on the display. We trust that the state does the job for us, and therefore we trust the petrol station staff. This authoritative backing does diminish suspicion, and in this sense standard weights and measures surely expand the social universe, as Simmel suggests. Whether this amounts to a moral advance is another matter.

In this story, there is of course a similarity between the scientific authority and the royal authority, as is the case for England in this period. For instance, Isaac Newton, spent many years as Master of the Mint in London, working precisely on the creation of coins of stable value and thereby on the authority of the state, whose sovereign's image was on the coins. In fact, the 1690s saw a monetary crisis prompted by the devaluation of coin by clippers. The political issues in recoinage were explicitly about the authority of the state. John Locke, who helped install Newton at the Mint, was of the opinion that the stake in recoinage 'was not simply an economic proposition, but social subversion, i.e., not just theft but treason.'[111] Locke argued that it was of the utmost importance that the state not admit any coins of substandard silver content; something which was common in all European states at the time. Many sovereigns decided to recoin with ever smaller amounts of silver, pocketing the rest for themselves, also in Scandinavia. But this created inflation, and in Locke's opinion, it sent the wrong signals. The state had to provide stable value and not descend to the level of clippers.

The trial of the Pyx at the London Mint established value of coins authoritatively through a ritual involving the presence of select gentlemen at a performance of an experiment involving precision measurement. The coins were then sent out into society where they acted as facilitators of exchange, guarding against fraud by providing trustworthy value. The scientific establishment of value in coinage and weights and measures was clearly a politics of consensus.[112]

CONCLUSION

There are remarkable similarities between Denmark-Sweden and Sweden-Finland. The metrological regimes were introduced into societies whose social development in terms of state organisation, property distribution of land, and exchange economies was very similar indeed. Both introduced administrational juggernauts in mid-17th century, which became the centres of fiscal-military states. Both set a premium on tax collection and both introduced standard weights and measures as a part of facilitating that collection. Both increased the tax level, both increasingly employed customs and excise, which again required standard weights and measures. Both supported the enclosures movement creating widespread ownership of land and enabled new methods of cultivation. Both had become money economies by 1800. The scientific institutions were founded at roughly the same time, and the metrological overlap between elite science and state requirements was the same in both. In both, a professionalisation of measurement involving oaths, monopolisation, exams, punishments, instructions, and routinisation took place.

The difference between the two countries is to be found in small details. Denmark-Norway remained absolutist through the entire period, Sweden-Finland had its Period of Liberty in the 18th century. Swedish science was more successful internationally, Copenhagen remaining a backwater. Swedish instrument making seems to have been much further advanced.

But if we compare the Scandinavian development with that of the southern European periphery, the contrast is quite stark, particularly with Greece. Greece, as part of the Ottoman empire, did not witness any of the developments mentioned above. There were no juggernaut administrations, no efficient fiscal regime along the lines described, no establishment of an efficient customs and excise, no surveying, no enclosures movement, no establishment of standard weights and measures, no academies of science, and no culture of instrument making. It can not surprise that the uptake of scientific ideas in Greece, and also in the Iberian peninsula, was of a completely different kind to the one in Scandinavia.

NOTES

[1] Tore Frängsmyr, J. L. Heilbron, Robin E. Rider (eds.), *The Quantifying Spirit in the 18th Century*, Berkeley: University of California Press, 1990

[2] M. Norton Wise (ed.), *Values of Precision*, Princeton: Princeton University Press, 1994

[3] Copenhagen Town Hall, Rådstuearkivet, Justerkamrets dokumenter 1696-1819.

[4] In fact, it is only really the development of this period which created the infrastructure in which land could be bought and sold. Land thus increasingly became a commodity from 1660 to 1800. In the late 17th century, the land tax was usually paid in kind, whereas in the late 18th century it was to be paid in money.

[5] Niels Morville, 'Om at bestemme Størrelsen af det Stykke Jord, som kan være tilstrækkeligt Vederlag for Tiendes Oppebørsel', *Skrifter som udi det Kiøbenhavnske Selskab af Lærdoms og Videnskabers Elskere ere fremlagte og oplæste*, New Series, 2 (1783), 25-48.

[6] Eli F. Heckscher, *An Economic History of Sweden*, Harvard University Press, Cambridge, Massachusetts, 1954; Ole Feldbæk, *Danmarks økonomiske historie 1500-1840*, Herning: systime, 1993; Sverre Bagge & Knut Mykland, *Norge i dansketiden*, Politikens Forlag, 1987

[7] For instance, the Danish law regarding land tax of 1662 stipulated a tax payable in grain, and the equivalent of 1791 stipulated a tax in money, cf. Birgit Løgstrup, *Jorddrot og offentlig administrator - Godsejerstyret inden for skatte- og udskrivningsvæsenet i det 18. århundrede*, Copenhagen: Rigsarkivet, 1983, 85-86

[8] Sten Lindroth, *A history of Uppsala University 1477-1977*, Uppsala: Almqvist & Wiksell, 1976; Svend Ellehøj & Leif Grane (eds.), *Københavns Universitet 1479-1979*, 14 vols., Copenhagen: C. E. Gads Forlag, 1979-1993

[9] Maija Kallinen, *Change and Stability: Natural Philosophy at the Academy of Turku, 1640-1713*, Helsinki: Finnish Historical Society, 1995

[10] Sten Lindroth, Kungl. Svenska Vetenskapsakademiens Historia 1739-1818, 2 vols., Uppsala: Almqvist & Wiksell, 1967; Tore Frängsmyr (ed.), *Science in Sweden - The Royal Academy of Sciences 1739-1989*, Canton, Mass.: Science History Publications, 1989. For the dichotomies, cf. Svante Lindqvist, 'The Spectacle of Science: An Experiment in 1744 Concerning the Aurora Borealis', *Configurations* (1) 1992, 57-94.

[11] Asger Lomholt, *Det Kongelige Danske Videnskabernes Selskab, 1742-1942; Samlinger til Selskabets Historie*, 5 vols., Copenhagen, 1942-1973

[12] Lindroth (note 8), vol.1, 'Almanacken', 823-867

[13] Svante Lindqvist, *Technology on Trial, The Introduction of Steam Power Technology into Sweden, 1715-1736*, Stockholm: Almqvist & Wiksell International, 1984

[14] For a very detailed analysis of Swedish scientists' travels and connections, along with the exchange networks of books, cf. Sverker Sörlin, *De lärdas republik - Om vetenskapens internationella tendenser*, Malmö: Liber-Hermods, 1994, esp. chs. 2-5

[15] Maurice Crosland, *Science in France in the Revolutionary Era Described by Thomas Bugge*, Cambridge, Massachusetts and London, 1968

[16] Svante Lindqvist, *Technology on trial*; and Feldbæk (note 6), 84

[17] Andreas Nissen, *Ole Rømer - Et Mindeskrift*, Fr. Bagges Kgl. Hofbogtrykkeri, København, 1944

[18] Professor Anders Celsii, 'Jämförelse emellan Den SWENSKA FOTEN och åtskillige utländske Mått', *Kungliga Svenska Vetenskaps-Akademiens Handlingar*, 1740, 216-218

[19] Jacob Faggot, 'TANKER om hwarjehanda Metallers och Träd-slags ändring uti storlek af Luftens Köld och Wärma: Lämpade til åtskilligt gagn uti Hushålning, Wetenskaper ock slögder,' *Kungliga Svenska Vetenskaps-Akademiens Handlingar*, 1740, 429-439; 'Om Tijotälning, eller Decimalers häfd i Bokhålleri ock Räkning, som rörer Måt, Mål Wigt och Mynt, utan rubning i de Wanlige inrättningar,' *Kungliga Svenska Vetenskaps-Akademiens Handlingar*, 1742, 49-57

[20] *Kungliga Svenska Vetenskaps-Akademiens Handlingar*, 1767, 158, 193, 209

[21] E Runeberg and H Nicander, cf. Register för Kungliga Svenska Vetenskaps-Akademiens Handlingar

[22] Z. Z. Plantin, 'Om ängers delning efter bördighet,' *Kungliga Svenska Vetenskaps-Akademiens Handlingar*, 1792, 239

[23] For a history of surveying in Sweden focussing on military projects and its connection with high science (*e.g.* the theories of the shape of the earth), cf. Sven Widmalm, *Mellan kartan och verkligheten - geodesi och kartläggning 1695-1860*, Institutionen för idé- och lärdomshistoria, Uppsala, 1990. A more balanced history than is given in this essay would in fact ensue if the work of Widmalm were to be better integrated.

[24] Hans R. Jenemann, *Die Waage des Chemikers*, Frankfurt am Main: DECHEMA, 1979; esp. 20-35

[25] A Berch, 'Jemförelse emellan Svenska och utländska vigter,' *Kungliga Svenska Vetenskaps-Akademiens Handlingar*, 1746, 274; 'Jemförelse emellan Svenska och Danska vigterna,' *Kungliga Svenska Vetenskaps-Akademiens Handlingar*, 1749, 223; 'Beskrifning på den Chinesiska vigten,' *Kungliga Svenska Vetenskaps-Akademiens Handlingar*, 1750, 210; J. Faggot, 'Beskrifning på en i Sverige bruklig lutvigt,' *Kungliga Svenska Vetenskaps-*

Akademiens Handlingar, 1743, 239; E. Runeberg, 'I Holland brukliga vikter,' *Kungliga Svenska Vetenskaps-Akademiens Handlingar*, 1759, 49. Also 1744, 222; 1775, 122; 1770, 259; 1772, 370; 1781, 217; 1781, 311

[26] Anders Berch, 'Jemförelse emellan Svenska kannemåttet och några utländska mått,' *Kungliga Svenska Vetenskaps-Akademiens Handlingar*, 1747, 274

[27] D. Ekström, 'Spannmåls-profvare, inrättad efter Svenskt mått och vikt,' *Kungliga Svenska Vetenskaps-Akademiens Handlingar*, 1753, 224; E. Runeberg, 'Spannmåls pröfning,' *Kungliga Svenska Vetenskaps-Akademiens Handlingar*, 1756, 276

[28] J. Faggot, 'Förbättring på dricksprofvare,' *Kungliga Svenska Vetenskaps-Akademiens Handlingar*, 1763, 45; idem., 'Om en accurat profvare för våta varor,' *Kungliga Svenska Vetenskaps-Akademiens Handlingar*, 1770, 255 & 270

[29] J. Faggot, 'Om runda kärls rymdmätning,' *Kungliga Svenska Vetenskaps-Akademiens Handlingar*, 1743, 200; Z. Z. Plantin, 'Coniska modeller til Svenska vigter och mål,' *Kungliga Svenska Vetenskaps-Akademiens Handlingar*, 1772, 370; also 1774, 156; 1782, 134

[30] Christen Hee, *Kort Underretning om det Danske Fodmaal, samt Forskiellen mellem Danske og Geographiske Mile*, 1770

[31] Niels Morville, 'Om gamle Danske Jorde- og Landgilds-Bøgers Translation og Landgilds-Speciers Beregning til Hartkorn', *Skrifter som udi det Kiøbenhavnske Selskab af Lærdoms og Videnskabers Elskere ere fremlagte og oplæste*, New Series, 5 (1799), 1-17

[32] Niels Morville, Om at bestemme Størrelsen af det Stykke Jord, som kan være tilstrækkelig Vederlag for Tiendes Oppebørsel. Afhandlet paa mathematisk Maade, *Skrifter som udi det Kiøbenhavnske Selskab af Lærdoms og Videnskabers Elskere ere fremlagte og oplæste*, New Series, 2 (1783), 25-48

[33] Niels Morville, *Geometrisk og oekonomisk Jorddeelings og Jordskiftnings-Lære til Nytte, saavel for dem, der forrette Udskiftnings-Forretninger, som og for dem, der lade sine Jorder udskifte af Fælledskab*, 1791

[34] Carl Deichmann, Om Maal og Vægt, som bruges ved Sølv-Verket (i Kongsberg), *Skrifter som udi det Kiøbenhavnske Selskab af Lærdoms og Videnskabers Elskere ere fremlagte og oplæste*, 11 (1777), 275-286

[35] F. H. Müller, Om Sølvets Prøvelse til Nytte for Mynte-Væsenet og Sølv-Handelen, *Skrifter som udi det Kiøbenhavnske Selskab af Lærdoms og Videnskabers Elskere ere fremlagte og oplæste*, New Series, 2 (1783), 153-173; F. H. Müller, Om Guld-Prøvens nøiagtigste Omgangsmaade, til Nytte for Mynt-Væsenet og Guldhandelen, *Skrifter som udi det Kiøbenhavnske Selskab af Lærdoms og Videnskabers Elskere ere fremlagte og oplæste*, New Series, 4 (1793), 1-28

[36] F. H. Müller, Om Maaden og Midlerne at undersøge, prøve og vurdere alle i Handelen forekommende Brændevine, i Hensigt saavel til deres Beskaffenhed, som Maal, med videre, *Skrifter som udi det Kiøbenhavnske Selskab af Lærdoms og Videnskabers Elskere ere fremlagte og oplæste*, New Series, 3 (1788), 202-219; F. H. Müller, Nøiere Oplysning og Forbedring vedkommende Brændeviinsprøveren og sammes Anvendelse, *Skrifter som udi det Kiøbenhavnske Selskab af Lærdoms og Videnskabers Elskere ere fremlagte og oplæste*, New Series, 5 (1799), 71-81

[37] David Kirby, *Northern Europe in the Early Modern Period - The Baltic World 1492-1772*, Longman, London and New York, 1990, 257

[38] Henrik Becker-Christensen, *Dansk toldhistorie II - Protektionisme og reformer - 1660-1814*, Toldhistorisk Selskab, København, 1988; Knud J. V. Jespersen, *Danmarks historie - Bind 3: Tiden 1648-1730*, Gyldendal, Copenhagen, 1989; Sven A. Nilsson, *De stora krigens tid - Om Sverige som militärstat och bondesamhälle; The Era of the Great Wars - Sweden*

as a Military State and its Agrarian Society, Uppsala: Almqvist & Wiksell International, 1990

[39] ibid.

[40] ibid.

[41] *Kammerkollegiets Historia*, Stockholm: Isaac Marcus Boktryckeri-Aktiebolag, 1941, esp. Nils Edén, 'Från Gustav Vasa till Karl XII:s död (1539-1718)', 1-194

[42] Birgit Løgstrup, *Jorddrot og offentlig administrator*, Copenhagen: Rigsarkivet/G. E. C. Gads Forlag, 1983

[43] Becker-Christensen (note 38); Jespersen (note 38); Heckscher (note 6); Göran Behre, Lars-Olof Larsson, Eva Österberg, *Sveriges historia 1521-1809 - Stormaktsdröm och småstatsrealiteter*, Esselte Studium AB, Stockholm 1985

[44] Karl Åmark, *Sveriges Statsfinanser 1719-1809*, Stockholm, P. A. Norstedt & Söners Förlag, 1961, 400-414

[45] ibid.

[46] Figures taken from Ole Feldbæk, *Danmarks økonomiske historie 1500-1840*, Herning: systime, 1993, 152

[47] *Svenska lantmäteriet 1628-1928*, Del I, Stockholm, 1928; Sven Widmalm, *Mellan kartan och verkligheten - Geodesi och kartläggning, 1695-1860*, Uppsala: Insitutionen för Idé- och lärdomshistoria, 1990

[48] H.T. Heering, 'Knud Thott og Forhistorien til Kristian V's Matrikul', in *Tidsskrift for Opmaalings- og Matrikulsvæsen* (13), 1932, issue no. 1, 1-19.

[49] For the tax system and the role of the local gentry in its management, cf. Løgstrup (note 42)

[50] The following is based on H. Juhlin Dannefelt, 'Skifte och delning av jord ur ekonomiskt synpunkt', and E. Williams, 'Skattläggningsväsendet och lantmätarna', both in *Svenska Lantmäteriet 1628-1928*, 2 vols, Stockholm: Sällskapet för utgivandet av lantmäteriets historia, 1928

[51] Feldbæk (note 6), esp. 109-115

[52] P. Nissen, 'Udskiftningstidens Landinspektører og Landmaalere', in *Tidsskrift for Opmaalings- og Matrikulsvæsen* (13), 1932, issue no. 1.

[53] Lomholt (note 11), vol. 5

[54] John Brewer, *Sinews of Power - War, Money and the English State 1688-1783*, London: Routledge, 1989

[55] Åmark (note 44), 400-414

[56] For Britain, cf. Brewer (note 54); For Denmark-Norway, cf: Becker-Christensen (note 38)

[57] Karl Åmark, *Spannmålshandel och spannmålspolitikk i Sverige 1719-1830 - Akademisk avhandling*, Stockholm, Isaac Marcus' Boktryckeri-Aktiebolag, 1915, 81-88

[58] Becker-Christensen (note 38), 457

[59] Cf. *Anordning om Maalere, Veyere, Vragere, Haufne-Fogder oc Faufnsettere*, Copenhagen: Bockenhoffer, 1683

[60] Becker-Christensen (note 38), 126

[61] Gunnar Pipping, 'Några drag ur det svenska mått- och justeringsväsendets historia,' *Dædalus - Tekniska Museets Årsbok*, 1968

[62] The following is based on John Svärdson, 'Lantmäteriteknik', and on H. Juhlin Dannefelt and E. Williams, all in *Svenska Lantmäteriet* (note 50)

[63] Svärdson, in *Svenska Lantmäteriet* (note 50), 8

[64] ibid., 25

[65] Witold Kula, *Measures and Men*, Princeton University Press, Princeton, New Jersey, 1986

[66] Pipping (note 61), 43-66, at 58

[67] Kirstine Meyer, 'Dansk Maal og Vægt fra Ole Rømers Tid til Meterloven', in *Beretning fra Meterudvalget om dets Virksomhed i Tiden fra dets Nedsættelse den 9. Juli 1907 indtil den 31. Marts 1914*, Copenhagen: J. H. Schultz, 1915, 57-91, esp. 57-70. Cf. also Arne Hægstad, *Mål og vægt i Danmark 1283-1983: den legale metrologi gennem 700 år*, Copenhagen, 1983

[68] ibid., 70-71

[69] Kula (note 65)

[70] Copenhagen Town Hall, Rådstuearkivet, Justerkamrets dokumenter 1696-1819. Also: Meyer (note 67), 74-76

[71] Svärdson (note 63), 52-64

[72] ibid., 50

[73] ibid., 86-87

[74] Gunnar Pipping, *The Chamber of Physics - Instruments in the History of Sciences Collections of the Royal Swedish Academy of Sciences, Stockholm*, 2. ed., Stockholm: Stiftelsen Observatoriekullen, 1991

[75] '6.2 Memorial til Riksens Ständers manufactur och Handels deput. ang. mathematiska Instrumentmakeriet, 29 martii 1739', Lantmäteristyrelsens expeditionsböcker B1:14 1731-1740, Lantmäteriverkets arkiv, Gävle

[76] Kula (note 65)

[77] Dan. Ekström, 'Beskrifning På en Spanmåls-profvare, inrättad efter Svänskt Mål och Vigt', *Kongl. Svenska VetenskapsAcademiens Handlingar för Månaderna Julius, Augustus, September, År 1753*, 224-241

[78] Københavns magistrat, MC600: 1696-1795: Justervæsen, Copenhagen Town Hall

[79] The alcohol excise was introduced in 1688, cf. Becker-Christensen (note 38), 256 and it became increasingly important for the state coffers throughout the 18th century.

[80] Franz Heinrich Müller, 'Om Maaden og Midlerne at undersøge, prøve og vurdere alle i Handelen forekommende Brændevine, i Hensigt til deres Beskaffenhed, som Maal, med videre', *Skrifter som udi det Kiøbenhavnske Selskab af Lærdoms og Videnskabers Elskere ere fremlagte og oplæste*, New Series, 3 (1788), 202-219; he completely changed the design later: 'Nøiere Oplysning og Forbedring vedkommende Brændeviinsprøveren og sammes Anvendelse', *Skrifter som udi det Kiøbenhavnske Selskab af Lærdoms og Videnskabers Elskere ere fremlagte og oplæste*, New Series, 5 (1799), 71-81.

[81] ibid. 203

[82] ibid., 205

[83] ibid., 209

[84] ibid., 203

[85] ibid., 212-214

[86] ibid., 204-205

[87] ibid., 212

[88] ibid., 216

[89] ibid., 217

[90] Keld Nielsen, *Hvordan Danmarkskortet kom til at ligne Danmark - Videnskabernes Selskabs opmåling 1762-1820*, Århus 1982, p.12.

[91] For Johan Ahl (1729-1795), cf. *Dansk Astronomi gennem firehundre år*, vols. 1-3, ed. by Claus Thykier, Copenhagen 1990, 191-192.

[92] Dan Ch. Christensen, 'Spying on Scientific Instruments - the Career of Jesper Bidstrup', in *Annals of Science*, 52 (1994).

[93] John Porteous, *Coins in history - A survey of coinage from the Reform of Diocletian to the Latin Monetary Union*, London: Weidenfeld and Nicolson, 1969, 15

[94] Sir John Craig, *The Mint - A History of the London Mint from A.D. 287 to 1948*, Cambridge University Press, Cambridge, 1953, 'Trial of the pyx', esp. 394-397

[95] Note items 19 and 28 in Ekström's list of instruments above; cf. also Michael A. Crawforth, *Weighing coins - English folding balances of the 18th and 19th centuries*, London: Cape Horn Trading Coy., 1979, who gives a good impression of the increasing distribution of precision balances towards the end of the 18th century. Maurice Daumas also covers precision balances on 221-227 of his *Scientific Instruments of the Seventeenth and Eighteenth Centuries and Their Makers*, London: Batsford, 1972.

[96] Salomon Haase, *Eröffnetes Geheimnus der praktischen Münzwissenschaft*, 1762; Salomon Haase, *Vollständiger Münzmeister und Münzwardein*, 1765; Johann Otto Ruperti, *Das Probiren, in so weit diese Wissenschaft zu dem Münzwesen nothwendig gehöret*, Braunschweig: 1765. Note that Ruperti's title refers to assaying as a science.

[97] Bjørn R. Rønning, *Den Kongelige Mynt 1628-1686-1806*, Norges Bank/J. W. Cappelens Forlag a.s., no place, 1986, 298-299

[98] ibid., 310-314

[99] The manuscript is entitled: 'Humble report on the Mint' (Underdån ödmiuk Relation om Myntet), and is reprinted in K.-A. Wallroth, 'Sveriges Mynt 1449-1917, bidrag till en svensk mynthistoria meddelade i myntdirektörens underdåniga ämbetsberättelser', *Numismatiska meddelanden utgivna av svenska numismatiska föreningen*, 12 (1918), 177-201

[100] ibid., 178

[101] e.g. ibid., 179

[102] A. Luschin von Ebengreuth, *Allgemeine Münzkunde und Geldgeschichte des Mittelalters und der neueren Zeit*, 2. ed., München & Berlin: R. Oldenbourg, 1926; Craig (note 94)

[103] This did not prevent sovereigns from attempting to make a profit from coinage, by issueing substandard coins, but state and trade always paid for this through the obstacles put in the way of smooth collection of tax and exchange of merchandise. Cf. Craig, passim.

[104] *Inventarium til den kongelige Mynt paa Kongsberg*, dated 11. januar 1806, Rigsarkivet (State archives), Copenhagen, Finanskollegiet, journalsager, jnr. 503, 1806; also printed in Bjørn R. Rønning, *Den Kongelige Mynt 1628-1686-1806*, Norges Bank/J. W. Cappelens Forlag a.s., no place, 1986, 323-329

[105] F. H. Müller, 'Sølvets Prøvelse til Nytte for Mynte-Væsenet og Sølv-Handelen', *Skrifter som udi det Kiøbenhavnske Selskab af Lærdoms og Videnskabers Elskere ere fremlagte og oplæste*, New Series, 2 (1783), 153-173; and 'Om Guldprøvens nøiagtigste Omgangsmaade til Nytte for Myntvæsenet og Guldhandelen', *Skrifter som udi det Kiøbenhavnske Selskab af Lærdoms og Videnskabers Elskere ere fremlagte og oplæste*, New Series, 4 (1793), 1-28

[106] Kula (note 65), 9

[107] ibid., 111-113, 127-146

[108] ibid., 16

[109] Peter Linebaugh, *The London hanged - crime and civil society in the eighteenth century*, Penguin Books, Harmondsworth, 1991, chapter 5: 'Socking, the Hogshead, and Excise'.

[110] Social anthropological research has thrown much doubt on the universal validity of the role of money, and emphasised the very local context in which money and barter take place. Cf. J[onathan] Perry & M[aurice] Bloch (eds.), *Money and the morality of exchange*, Cambridge: Cambridge University Press, 1989

[111] Constantine George Caffentzis, *Clipped coins, abused words and civil government - John Locke's philosophy of money*, New York: Autonomedia, 1989, p. 35. In the footnote Caffentzis quotes Locke: '[T]he use and end of the public stamp is only to be a guard and voucher of the quantity of silver which men contract for; and the injury done to the public faith, in this point, is that which in clipping and false coining heightens the robbery into treason.' John Locke, 'Further Considerations', in *The Works of John Locke in Ten Volumes* (London, 1823) reprinted by Scientia Verlag Aalen, Germany, vol. 5, 144. Locke wrote this in 1696.

[112] For science and the politics of consensus, cf. Steven Shapin & Simon Schaffer, *Leviathan and the Airpump*, Cambridge: Cambridge University Press, 1985

Printed in the United States
85705LV00001B/43-44/A